Unlocking Alpha
The Rise of Niche VC

Simon Lancaster
with
Sabrina Paseman

Orion Press
15466 Los Gatos Blvd.
Ste. 109239
Los Gatos, CA 95032

ISBN-13: 979-8-9998196-0-4 (hardcover)
ISBN-13: 979-8-9998196-1-1 (softcover)
ISBN-13: 979-8-9998196-2-8 (ebook)

Table of Contents

Foreword.. 1

Acknowledgments 3

Introduction: Chasing New Frontiers 5

Part I: The Value of Niche in Venture 9

Chapter 1: The New Wave of Venture Is Niche................................. 11

Chapter 2: The Data Favors Specialization 17

Part II: The Past and Present of Venture 25

Chapter 3: The Golden Years... 27

Chapter 4: The Lost Years 39

Chapter 5: Rethinking Modern Venture 45

Chapter 6: The Rise of the Niche VC 57

Part III: A How-To Guide for Niche 63

Chapter 7: How to Pick a High-Performing Niche VC 65

Chapter 8: Designing Your Own Niche Strategy 75

Chapter 9: Executing on Your Niche Strategy 91

Chapter 10: Building a Lasting Firm 97

Conclusion: The Future of Venture 101

Glossary.. 103

Foreword

The world is at a breaking point.

Markets feel broken. Capital has grown complacent. Institutions are flooded with noise yet starving for signal. Alpha—real, meaningful, world-shifting alpha—is harder to find than ever. It's no surprise: The industry has over-rotated into consensus, spreadsheets, and brand-name herding.

But in every cycle there are a few who don't just survive—they see the next wave before anyone else. And they bet on it early, quietly, and with relentless conviction. These are the niche VCs, and they're not a trend. They're the system-critical minority we need now more than ever. Why? Because they are guided by something that can't be spreadsheet-modeled or mimicked: intuition—that deeply wired pattern recognition that comes only from lived experience and real mastery. Einstein once called it the "only real valuable thing," and in our experience, it's what separates noise from signal, especially at the edge of what's next. That's why niche VCs matter.

At Allocator One, we've analyzed over eight hundred emerging fund managers. Less than 3 percent make it through. But when they do, they often look like this: relentless, sharp, respected in their field, and able to see around corners with a clarity that others can't even imitate.

This book isn't just a guide to niche VC. It's a blueprint for the future of capital. If you want to understand where the real returns—and the real impact—will come from in the next decade, don't look to the crowd. Look to the niche. And start right here.

Felix Staeritz & Michael Ströck
Founding Partners, Allocator One

Acknowledgments

This book would not exist without the many people who believed in us, guided us, and endured our endless questions along the way.

To **Chris Kim**, our first and forever GP mentor, who was always just far enough ahead for us to follow his path while urging us to pave our own. You helped conceive this book, shaped its title, and gave feedback at any hour with a calm, rational voice that's guided us through countless deals and LP conversations.

To **Lucas Whipple**, for believing in the "crazy" concept of our niche VC firm when no one else did, becoming both our first LP and first venture partner, and lending deep technical diligence in industrial and software automation.

To **Benedikt Langer**, our first "LP insider," thank you for giving us the candid insights we desperately needed in the early days—back when we were still figuring out what an LP or a "family office" even was.

To **Chris Harvey**, our tireless legal counsel, who took calls at all hours; navigated CFIUS, QSBS, fund formation, and complex deal structures with precision; and became a de facto member of our team.

To **Chris Yeh**, whose work with Blitzscaling inspired one of our first angel investments and who brought us along on speaking tours, showing by example what collaboration and generosity in venture look like. Your advice on writing and publishing has been invaluable.

To **Paul Blanchard**, for PR support (and PR therapy!) and for inspiring this book through our countless conversations and through your own book *Fast PR.*

To **Winter Mead** at Coolwater Capital, for pushing us toward a true institutional Fund I instead of a small proof of concept.

To all our **first-closing LPs**, and especially **Allocator One**, our first anchor—your early belief gave us the momentum to reach final close and made both this firm and this book possible.

To all **twenty-three incredible founders** we've backed—especially **Firas Khalifeh**, who built alongside us in the chaotic pre-fund days— thank you for your trust and hustle.

And finally, to our **spouses, parents, and children**, for their patience, sacrifices, and love during years of late nights, early mornings, and the dual challenge of being emerging GPs and authors.

Introduction:
Chasing New Frontiers

Before satellites and GPS, the world's greatest explorers relied on cartographers to shape their understanding of the unknown. These early mapmakers didn't just sketch coastlines—they climbed hills to see what others couldn't. They studied wind patterns, trade routes, terrain, and tides. They knew how to read a landscape before anyone else even thought to explore it. In a world of speculation and rumor, cartographers were trusted not because they guessed but because they knew where to stand and what to look for.

Today's venture capital landscape mirrors that same age of exploration. Investors from around the globe are flooding into the industry hoping to strike gold, but few know where to dig. They see big wins—an OpenAI, a Stripe, a Figma—and rush in, hoping to catch the next one. But by the time a space looks hot, it's already crowded. Most capital ends up chasing consensus ideas at consensus prices. That's not how you find alpha—"a measure of an investment's performance that indicates its ability to generate returns in excess of its benchmark."[1]

True alpha comes from seeing the opportunity before the map is drawn. Niche VCs are the new cartographers—specialists who've spent years mastering a terrain most others overlook. They climb the hills, spot the signals, and chart the frontiers of innovation before the rest of the market catches on. Over the last seventy years, each new wave of breakthrough venture returns has come not from generalist spray-and-pray investing but from this kind of focused mastery. Semiconductors in the 1950s. Biotech in the '70s. Web services in the early 2000s. Mobile.

---

[1] James Chen, "Alpha: Its Meaning in Investing, with Examples," *Investopedia*, updated February 23, 2024, https://www.investopedia.com/terms/a/alpha.asp.

Software as a service (SaaS). AI. And now, as capital dries up and the market matures, the pendulum is swinging back to deep sector-specific insight. The niche venture capital (VC) is no longer an exception—it's the new frontier.

This book is for any venture capitalist in search of exponentially better returns. In part I, I cover the characteristics of outstanding niche VCs and why firms that specialize are best positioned to find alpha in today's crowded market. Part II traces the history of venture and how its pioneers built mastery, focus, and network to consistently outperform. Part III offers a practical guide for designing and executing a niche strategy of your own, whether you're a limited partner (LP) evaluating funds or a general partner (GP) building one.

As a cofounder and GP at Omni Ventures, a niche VC specializing in pre-seed investments in manufacturing-tech start-ups, I've walked every step of the path outlined in this book. This journey has not been an easy one. In the face of breakups with founders, mind-blowing lawsuits, and the evaporation of capital that followed the most recent zero-interest-rate policy (ZIRP) era, the key to perseverance has been the singular factor most successful GPs and founders must share to survive: ridiculous levels of relentless grit.

This is a hard game, but an endlessly rewarding one for anyone who wants to influence the future while reaping the financial rewards that come with investing in the right companies at the right time. Early-stage venture is a multiplayer, non-zero-sum game; small VCs are stronger when we work together in this evolving field. As someone who has benefitted massively in the past from the support of colleagues in my network during tough times, I'm a firm believer in paying it forward. A constant goal of mine is to continue expanding the network of VC superstars I've built by aligning forces with those around me who are in the same boat. This book is my contribution to help anyone looking to level up or break through walls standing in the way of the colossal returns they hope to see.

I turned down a nearly seven-figure salary offer from Meta to cofound my own VC, and the reasons were twofold. The first was my desire to work closely with founders, who are some of the most passionate, hardworking people solving the world's biggest pain points. To surround myself with these people has been an irreplicable privilege. The second stemmed from my fascination with the diverging relationship between GDP growth and population growth. We are currently in the midst of the Fourth Industrial Revolution. The first arrived with the invention of the steam engine. The second came with mass production. The third came about with computers and microprocessors that enabled society to automate work in a variety of sectors. Throughout each of these new technological waves, GDP and population growth have remained closely correlated. As the rise of robotics and AI ushers in the fourth wave, the potential output of a single individual has skyrocketed. This potential for widening ripples of impact and profit in the world of VC intrigues me. I'm determined not only to scale my impact and get a piece of the action for myself in the process but to open the doors necessary to help you do the same.

Part I:
The Value of Niche
in Venture

Chapter 1:
The New Wave of Venture Is Niche

A Look Ahead (The TL;DR)

- At this point in history, the VCs best positioned to generate alpha are those that specialize in a niche they can dominate.
- Great niche venture firms share three characteristics highly correlated with alpha: mastery, focus, and network.
- Narrowing a fund's focus to serve a particular niche gives these firms a unique perspective that enables them to spot obscure opportunities before anyone else.

Back in the 1990s, as venture firms ballooned into empires, one new firm defied the trend. Benchmark Capital, now known for zigging when others zag, opened its doors in '95 with a radical philosophy: "Venture capital doesn't scale." Rather than raising billion-dollar funds or chasing every hot deal that popped up, Benchmark kept its funds small, its team lean, and its focus on early-stage bets that it understood deeply. Critics scoffed at the time. How would remaining tiny enable them to compete with VC titans? Regardless, Benchmark went ahead and backed eBay with a mere $6.7 million investment. When eBay finally went public, that single bet drove Benchmark's fund to an astounding forty-seven times return at the fund level.

Over the following years, this "tiny" firm repeatedly hit it big as an early investor in Twitter, Uber, and Instagram—all with modestly sized funds. Benchmark's conviction that "bigger is not better—*better* is better" paid off. Its story has become a legend in Silicon Valley. By choosing *not* to build a sprawling assets-under-management (AUM) empire, Benchmark became one of VC's most successful franchises. This lesson sets the stage for the premise of this book: At this point in venture history, its most impactful players are likely to be those who stay focused and agile. In VC, as in business, sometimes less is truly more.

Some might consider Benchmark to be a controversial example of niche VC given that its fund is rooted in a laser-focused generalist strategy rather than a sector-specific one. Yet the framework I introduce in this book defines niche not by *what* you invest in but by *how* you run the firm. Benchmark's intentionally small fund size, unwavering brand, and almost monk-like discipline set it apart from the sprawling mega-fund crowd—and make it a textbook example under this new lens.

What Is Niche VC?

In the pages ahead, I lay out a flywheel with three spokes that spin up outsized alpha. Master these three elements and a firm earns the right to call itself truly niche:

1. **Mastery:** Hard-won domain insight that enables a firm to spot signal while the herd is still guessing at the noise.
2. **Focus:** Disciplined, thesis-driven filtering that passes on 99 percent of deals so the fund can hammer the 1 percent that truly matter.
3. **Network:** A magnetic brand, community, and reputation that pulls founders, coinvestors, and strategic partners into a GP's orbit before a term sheet ever lands.

Rather than existing in a binary, each aspect of this trifecta lies on a spectrum. A GP may be high in mastery, for instance, but low on focus. By this metric, Benchmark would score low on mastery but very high on focus and network. The point is that some amount of each of these three ingredients is necessary for driving alpha. In an ideal scenario, the following interplay occurs:

→ **Mastery** begets sharper **focus**.

→ **Focus** amplifies brand signals and deepens the fund's **network**.

→ The **network**, in turn, expands the surface area where **mastery** can compound.

$$\text{Mastery} \times \text{Focus} \times \text{Network} = \text{Alpha}$$

There is no single template for designing a niche fund. We'll explore a wide variety of possibilities later in this book, but for now, here are some common ways GPs narrow their funds.

Table 1. Lenses Through Which GPs Narrow Their Funds

Lens	Sample Focus
Sector	Biotech (IndieBio) Climate/transport (Trucks VC) Consumer tech (Upside Ventures)
Stage	Pre-seed only (Afore & Notation) Series A only (Emergence Capital)
Geo	Ukrainian defense tech fund (D3 VC) Israeli cyber (YL Ventures)
Worldview	Focused on backing local founders in emerging markets (Don't Quit Ventures) Physics-driven AI Winner-take-all markets (Blitzscaling Ventures)

Why Niche VC Outperforms

In their best-selling book *Zero to One*, Peter Thiel and Blake Masters explained the power of dominating and scaling within a niche market. They argued that true progress comes from vertical innovation—creating a fundamentally new piece of tech rather than improving on products people already use. Rather than competing within a crowded market, great start-ups build monopolies around a secret most people don't know or believe is possible. They provide entirely new solutions none of us knew we needed.

> *Competition is for losers.*
> —Peter Thiel and Blake Masters

By extension, VCs who dominate mastery in promising niches provide investors with what Miles Deamer from Crossover VC refers to as a *hilltop advantage*. This concept is simple but powerful: It describes a master specialist VC's ability to survey the landscape from a higher perch, spotting connections and white space that flatland generalists overlook. In other words, they possess a perspective that lets them see into the distance and envision ideas others can't even conceive of.

In my own career, I've been fortunate to climb a few of these "hills," from consumer electronics and materials science to manufacturing software, supply chain, and even biotech. Each vantage point offered a new way to connect dots across disciplines, revealing opportunities and solutions invisible from any single one of those domains alone. It's how I've accumulated over fifty patents—by fusing disparate ideas into something fundamentally new. That's mastery in action.

The Edge Afforded by Mastery

Unique domain insight gleaned from experience primes a fund for alpha in a number of ways.

Signal Before Consensus

A focused strategy allows GPs to see signal long before consensus exists. Rather than chasing crowded deals with frothy valuations, niche VCs build deep pattern recognition. They know what truly matters, what will scale, and what's just noise. This gives them the confidence to take calculated risks that generalist VCs might be too timid or too uninformed to take. They understand the market dynamics, customer pain points, and regulatory hurdles of their niche better than anyone, letting them underwrite opportunities with far more conviction and often far less competition.

Repeatable Strategy

Niche venture also creates a repeatable strategy for diving deep into new areas, gearing the best minds toward unexplored corners of the market. They're not just allocating capital; they're compounding insight, building mental playbooks, and spotting the next frontier before the broader market even knows to look. Their specialization enables them to consistently deploy capital into concentrated, high-conviction bets without distorting the market—something large, generalist funds structurally struggle to do.

Founder Access

Perhaps most importantly, the best founders actively seek out niche VCs because they want partners who understand what they're building, not just faceless capital chasing the next theme. In high-stakes, complex industries like spacetech, manufacturing tech, or biotech, this becomes a critical differentiator. Specialized VCs come to the table with a number of aces up their sleeve: They help start-ups scale faster through hard-won operational knowledge, high-trust customer introductions, derisked supply chains, strategic hiring, and product development advice. They grasp industry pain points that can't be learned from books or spreadsheets alone because they've lived them.

Common Concerns About Niche Venture

Isn't the total addressable market (TAM) too small with niche funds?

The TAM is the estimated market demand for a certain product or service, which can be used to size the implied revenue opportunity. Your market will never be too small as long as the frontier of your niche is expanding. Use of AI in the drug-design corner of the pharmaceutical industry, for instance, looked microscopic a decade ago. Today it's a multibillion-dollar scramble.

Won't our VC miss the next big thing if we narrow our focus to specialize?

Not necessarily. It's important not to equate breadth with insight. Specialists routinely identify paradigm shifts first because they're embedded right where those shifts start.

You don't have to believe me about the potential of specialization in today's VC scene, however. The rise of niche as venture's newest frontier is more than a theoretical concept; it's an evidence-based phenomenon supported by years of recent data. We'll take a look at the numbers next.

Chapter 2:
The Data Favors Specialization

A Look Ahead (The TL;DR)

- When it comes to finding alpha in venture, the data shows specialists are systematically beating generalists.
- Outsized returns concentrate around the most disciplined, focused, founder-embedded GPs.
- Specialization is statistically rewarded up to the point that a check's size stretches beyond what that edge can underwrite.
- Specialization is a common thread among VCs that have managed to thrive throughout past extinction events.
- Opportunity cost is high for emerging GPs that go all in on managing a niche fund. The generalist approach is easier, which is why it's now the status quo.

Biotech VC firm Flagship Pioneering has originated over one hundred companies since its founding in 2000, resulting in more than $70 billion in aggregate value.[2] Its model, predicated on extreme specialization, has sourced talent internally around platform technologies like synthetic

---

[2] Chris Witkowsky, "Longtime Flagship GP David Berry Seeks $300m for New Firm," *Venture Capital Journal*, January 10, 2024, https://www.venturecapitaljournal.com/longtime-flagship-gp-david-berry-seeks-300m-for-new-firm/.

biology and programmable mRNA years before Wall Street catches wind of the buzz. Flagship's 2012 Fund IV, in particular, has proven to the VC world just how lucrative a deep-biotech focus can be. The $700 million vehicle is currently on track to return nine times the invested capital, the best single-fund multiple ever recorded in the sector, all thanks to the hilltop advantage created through its laser-focused niche.

While Flagship's returns have been extraordinary, the stats around its outstanding performance illustrate an emerging pattern. Independent datasets keep converging on the same conclusion: Specialists are systematically beating generalists in VC. The alpha is clear.

Niche-VC Case Studies

In venture, the delta between "good" and "great" is vast. It's not enough simply to allocate to niche; it's about identifying top-performing GPs who consistently deliver outlier returns. These GPs, typically running sub-$50-million specialist funds, do not outperform by chance. They combine domain mastery, founder access, and discipline to create return profiles that are statistically distinct from those of their peers.

Table 2. A Comparison of Fund Categories by Internal Rate of Return (IRR), Total Value to Paid-In Capital (TVPI), and Distributed to Paid-In Capital (DPI), Expressed in Multiples of Initial Investment (×)

Fund Category	Net IRR Range	TVPI (Median)	DPI (Median)
Top-performing <$50M niche funds	28%–33%	2.8×–3.5×	1.3×–1.6×
Top quartile <$250M funds	22%–28%	2.2×–2.6×	1.0×–1.3×
Top quartile generalist >$250M mega-funds	17%–21%	1.9×–2.2×	0.8×–1.1×

Note: Data in column 1 sourced from PitchBook 2023 (1,400 funds analyzed [2015–2020 vintages]), Cambridge Associates Private Equity and VC Benchmarking Data (2001–2020), and Allocator One internal benchmark (800+ early-stage funds across the US, Europe, Latin America, MENA, and Asia).

While a niche model can never guarantee alpha, table 2 displays how outsized returns concentrate around the most disciplined, focused, founder-embedded GPs. These managers deliver performance metrics 10 to 15 percentage points above their peers in net IRR and generate DPI earlier without relying on mega-round markups.

Case Study 1: PitchBook 2023 Analysis

Research firm and financial data provider PitchBook performed an analysis of 1,400 venture funds in 2023. The pattern of niche outperformance held steady throughout its results. In sub-$250-million funds, the hunting grounds of most emerging GPs, specialist funds were the clear winners on both net IRR and TVPI, with the gap widening as vintages seasoned. For funds under $250 million, specialists led the way on both IRR and TVPI across all cohorts (see figure 1). Funds over $250 million with a generalist model grabbed an approximately 5 percentage point IRR edge in the 2015–2020 bull market (see figure 2), while TVPI remained neck and neck.

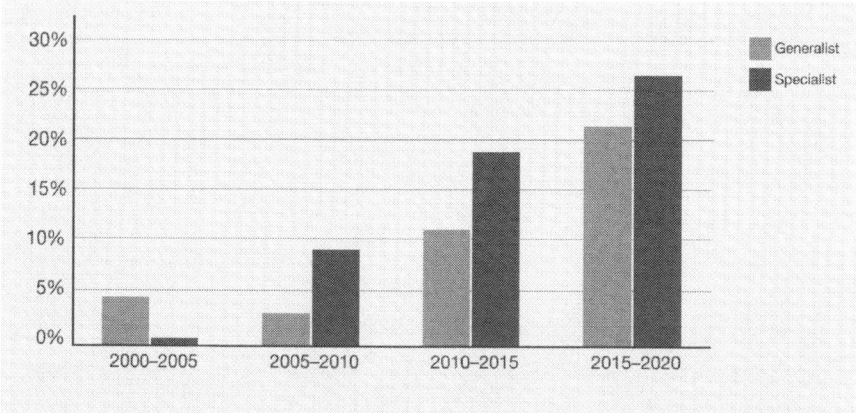

Figure 1. Performance of VC Funds Under $250 Million by Vintage Cohort and Style

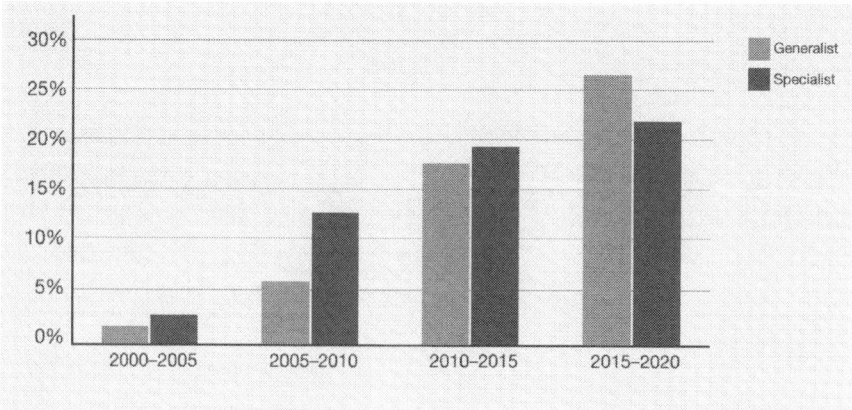

Figure 2. Performance of VC Funds over $250 Million by Vintage Cohort and Style

Case Study 2: Xi Han's 2009 Paper on Specialization

The 2009 paper "The Specialization Choices and Performance of Venture Capital Funds" by Xi Han, an investor at Tao Capital, further supports the argument that focus—not fund size—drives alpha. Working with a monster dataset of 1,586 US-based VC funds covering 64,168 investments, Han showed that specialization is a continuum: Some funds are razor focused, while others spray capital across industries, stages, and states. The degree of focus tightens when a GP wields deep domain expertise and loosens as funds swell in size or GPs grow more risk averse. These mechanics explain why many breakout niche franchises eventually dilute their edge. Crucially, Han's regressions revealed that industry and stage concentration materially boost initial public offering (IPO) and mergers and acquisitions (M&A) hit rates, whereas geographic focus offers no consistent advantage. Bottom line: Specialization is statistically rewarded—*until a check's size stretches beyond what that edge can underwrite.*

Our Most Painful Lessons:
Overcapitalization Is Real

Too much money too early can kill a start-up. We've seen it firsthand. When a founder raises ahead of their readiness, it often ends in misalignment, burn, and disappointment. Help them pace, not sprint.

Case Study 3: Cambridge Associates 2010 Analysis

From 2001 to 2010, global investment firm Cambridge Associates analyzed 3,700 US buyout and growth-equity deals. Its results showed that specialist investments delivered a 23.2 percent gross IRR and 2.2 multiple on invested capital (MOIC, expressed as ×), versus 17.5 percent and 1.9× for generalists—a 570-basis-point spread and 30 percent uplift in multiples. Just 14 to 25 percent of specialist capital ended up below cost, compared with 28 to 31 percent for generalists, while 40 to 52 percent of specialist dollars landed in greater than 2× winners.

Case Study 4: Previous Extinction Events

Past catastrophic market crashes provide further insight into the resilience of specialized funds. The extinction events of 2002–2003, 2009, and 2023 each followed a bubble (dot-com, credit/cleantech, ZIRP) that had rewarded shallow, trend-chasing strategies. After the dot-com crash, only 108 US VC funds raised money in 2002, a 95 percent plunge from 2001. Twenty-six other funds ended up returning $5 billion to LPs, the worst fundraising year since 1981.[3] Data from the National Venture Capital Association (NVCA) and VC fund Thomson Reuters showed fundraising collapsing from $106.9 billion in 2000 to $6.9 billion in 2002—a drop of roughly 93 percent.[4] The common thread among VCs that survived these drastic implosions? You guessed it: specialization. During each era that followed an extinction-level crash, firms that endured tended to have deep sector insight, proprietary deal flow, and disciplined fund sizes.

The Challenges of Niche VC

By now you may be wondering, if specialization is such a promising path to alpha, why aren't most VCs niche-focused already? The answer, simply put, is that starting a specialized firm poised to provide unique value is very hard to do. Opportunity cost is a serious consideration for emerging niche GPs, and most people in the venture world don't fully appreciate this. A person must essentially put their life on hold and go all in on becoming self-employed—similar to a start-up founder—to get their fund up and running with a strategy that is both effective and sustainable. I've seen GPs walk away from six-figure salary packages with full health benefits and postpone significant milestones like buying a house or having a baby. It's worth noting that many professionals who achieve mastery over a specific sector come to venture without the

---

[3] Associated Press, "VC Funding Slows to a Trickle," *Wired*, February 11, 2003, https://www.wired.com/2003/02/vc-funding-slows-to-a-trickle/.

[4] Eric Hazard, "Venture Funds See Belt Tightening in 2002," *PLANSPONSOR*, February 11, 2003, https://www.plansponsor.com/venture-funds-see-belt-tightening-in-2002/.

advantage of an existing LP network and must build their own from scratch, adding more risk to an already risky career move.

Without sufficient capital, a supportive spouse, a dedicated team, years of sector-specific experience, and an endless passion for cutting-edge tech, getting a niche-VC firm off the ground becomes impossible. For many aspiring GPs, the initial level of sacrifice required is ultimately too much to bear. Additionally, staying self-disciplined at the frontier of this hype-driven industry takes an insane amount of determination and grit. Even GPs who make it past the hurdles of building a fund can easily fail by widening and diluting their thesis over time.

Our Most Painful Lessons:
How Much Runway Do You Really Have?

Before you raise a fund, ask yourself, "Can I survive twenty-four months without a salary?" Add travel costs, living expenses, and a lot of emotional stamina. If the answer is no, wait until the answer is yes. This is a burn-first business.

LPs seeking alpha must navigate the challenge of identifying GPs who, like great founders, are intrinsically motivated to build. The high opportunity cost of forming a fund with a specialized focus should serve as a clue in their search. If talented experts are sacrificing assets like great health insurance and financial security to go all in on becoming a GP of a niche VC, they may know something others don't about the potential of their firm that may be worth looking into. These are the GPs most likely to win big as generalized mega-funds continue to dig up meager returns in an overcrowded market. Following the insights of specialized VCs with aligned interests can give interested LPs an outrageously unfair advantage.

The subsequent chapters of this book cover the history of venture as it relates to specialization and offer strategies for identifying and building niche funds that outperform.

Part II:
The Past and Present of
Venture

Chapter 3:
The Golden Years

A Look Ahead (The TL;DR)

- Venture capital has always been about investing in new frontiers of innovation. Understanding VC's history provides insight into its current commoditized state.
- The venture industry stemmed from angel investing in the mid-1900s. During the '50s, Arthur Rock and the Traitorous Eight brought VC to Silicon Valley.
- Venture matured over the years through frontier-focused deals around biotech, computing, and web-based technologies.
- Mega-rounds became commonplace in VC, as the high scalability of SaaS tech made it possible for start-ups to grow faster than ever before.
- As GPs focused more on SaaS metrics and less on frontier-focused innovation, differentiation among VCs became scarce and a renewed hunger for specialization began to grow.

In the high-risk, high-reward world of venture, failure isn't a reflection of incompetence but the price of discovery. It's the only game where batting one hundred can turn players into legends. While 90 percent of start-ups die quietly, it takes only one success story to birth an entirely new industry. In 1976 when Kleiner Perkins invested in Genentech, biotechnology didn't yet exist as an industry. Its investors were gambling on unproven science pushing boundaries at the edge of the map where the odds of success were slim. Genentech's eventual success not only founded the biotech industry but also led to a $2.1 billion acquisition by Roche, one of the world's largest pharmaceutical and diagnostics companies. It was a bold bet that went against the grain. Investors built their own conviction before consensus had been reached. Back then, hitting alpha with moonshots like these was venture's main purpose.

From the days of its inception, venture capital has been a practice of investing in new frontiers. As the industry flourished over the second half of the twentieth century, VCs funded deep-tech start-ups made up of pioneering visionaries with highly specialized skills. Each home-run investment signified a cutting-edge niche with explosive returns. By scanning VC's history from a bird's-eye view, we can take note of what worked in the past and begin to understand what's watering down venture today.

Angel Investing: Venture's Precursor

Long before venture capital established itself as a formal industry, angel investors were the well start-ups drew from for funding. As early as the 1900s, industrialists, bankers, and family dynasties like the Rockefellers used their own money to back start-ups driving new technology. Some invested based on their personal relationships, while others had a strong intuition for innovation, but few operated according to any structured process. Angel investing was sporadic, unscalable, and difficult to navigate, driven by wealthy risk-takers who could tolerate failure.

The first angel to institutionalize venture was Harvard professor and World War II logistics officer Georges Doriot. In 1946, Doriot launched the American Research and Development Corporation (ARDC), the first VC firm to make investments using outside capital. ARDC went on to back the Digital Equipment Corporation with $70,000, recognizing its potential and sharing its founder's vision. This resulted in more than a 5,000× return of over $350 million upon exit, proving to other angels and prospective investors that a standard, structured approach could help them strike gold too. This is when angel investing, a personal and scattered process, blossomed into the repeatable, professionally managed world of VC.

This evolution of institutionalization explains venture's distinctly high-risk, founder-driven culture. Early VCs were former angels, for the most part, who poured their risk tolerance and relationship-driven ethos into their firms while adding exit planning, due diligence, and board seats along the way. This allowed them to bring a degree of order to the unpredictability of a game with no guarantees.

Venture's First Big Break

Throughout most of the 1950s, America's wealthiest families began forming "risk capital" firms that aimed for explosive growth but still lacked a common structure. The deals they made were private and informal, focusing on Cold War–driven innovation in defense, aerospace, and electronics. Venture's true breakthrough didn't arrive until the late '50s, when a team of eight young and rebellious engineers made the choice to walk away from an oppressive employer.

The Traitorous Eight had been working for William Shockley, coinventor of the transistor, who founded Shockley Semiconductor in Mountain View, California. While Shockley inspired awe with his brilliance and showmanship, he was an erratic tyrant who taunted his employees, staged public firings, and belittled researchers. In 1957, eight of those employees, led by twenty-nine-year-old Robert Noyce, revolted and quit, determined to take their highly specialized skills elsewhere. The only question at hand was how to fund their work. They were soon introduced to Arthur Rock, a reserved New York banker who convinced the group to start its own company. Rock also convinced Fairchild Camera and Instrument, which provided research and development for flash photography equipment on the East Coast, to back the group with the funding it needed. The resulting company, Fairchild Semiconductor, was born and soon revolutionized chip manufacturing through its invention of the planar process, a key element in the manufacture of modern microchips.

Over time, Fairchild became a breeding ground for founders of deep-tech start-ups. Its alumni left to create companies such as Intel, AMD, National Semiconductor, and Kleiner Perkins. Before that point, venture capitalists focused largely on industrial and medical innovations, but the Fairchild deal cemented technology as a VC magnet. Rather than pouring money into companies with government contracts or inventors with prototypes, venture capitalists pivoted to rebellious young techies, visionaries, and groundbreaking founders in need of capital. Arthur Rock, who went on to fund Apple and help create Sequoia Capital, had found success as a VC by backing not an idea but a team. A core principle of Silicon Valley venture capital was born: Bet on people, not just products.

Our Most Painful Lessons:
Support the Rising Stars Early

Don't wait for the deck. Help founders before they even have a company. The loyalty you build before they're successful is what gets you into their best deals when they are.

Venture Grows Up

What began as an elite club of wealthy families and former bankers solidified into a professional financial industry backed by outside capital during the '70s. One of VC's household names at the time was Kleiner Perkins, a firm known for its deep-tech focus and hands-on involvement, which funded early Silicon Valley superstars such as Genentech and Sun Microsystems. There was also Sequoia Capital, founded by former Fairchild and National Semiconductor executive Don Valentine. He went on to fund Apple in 1977 when Steve Jobs was just twenty-one years old. Another major player was Mayfield Fund, known for its investments in semiconductors and computing infrastructure, which went on to back Tandem Computers.

A crucial game changer in 1974 came in the form of the Employee Retirement Income Security Act (ERISA), which allowed pension fund managers to invest in high-risk assets they had previously been banned from. Billions in LP capital were unleashed through VC firms, cementing venture as a legitimate asset class. This era set the stage for the first real boom during the '80s, when firms scaled the industry's previously formed foundations into a recognizable financial powerhouse. Personal computing and biotech companies began producing billion-dollar outcomes, thrusting VC into the spotlight of mainstream press.

Interest in personal computing, emerging software, and hardware start-ups skyrocketed following the success of Apple's 1980 IPO. In the same year, Genentech's IPO helped launch the biotech sector, inspiring the creation of similar companies. From that point onward, IPOs became a primary liquidity event for VCs, sparking feverish frenzies on Wall Street and causing tech companies to go public earlier with higher valuations. Breakout wins included Lotus Development Corp., Compaq, Amgen, and Sun Microsystems. By the mid-'80s, funds of $100–$300 million were common, and certain VCs had multiple greater than 10× exits under their belts. As these firms developed reputations, LPs began tracking their performance, while founders began approaching them as strategic mentors in addition to capital allocators.

Throughout venture's initial maturation period, smart VCs struck gold by focusing on founders who were more technically skilled than business savvy. Nearly every successful deal backed deep tech that required considerable expertise to produce. The GPs themselves often wore both hats, supporting founders in business and technical specialization. From this standpoint, venture capital began as an asset class invested in niche technology, where generating alpha hinged on a GP's ability to identify potential in products the world had yet to conceive of. By zeroing in on each new frontier, eyes glued to the horizon, VC became known worldwide for its ability to make risk-tolerant investors obscenely rich.

Venture's initial ties to niche tech launched the industry as a forum where relatively few risk-takers could achieve the American Dream through specialization. Start-ups were idiosyncratic bets that revolved around unique products, markets, and business models. This all changed as the turn of the century approached, however, as the dot-com gold rush flooded the market with investors who made software their primary focus.

The Dot-Com Boom

In 1991, the World Wide Web went live to the public for the first time. By the middle of the decade, the VC scene was teeming with feverish excitement over the explosive growth of the internet and all the ways it could change the world. This frenzy was initially sparked by Netscape's early IPO in 1995 as the company went public just sixteen months after its founding. The IPO, priced at $28 per share, saw its stock price soar to $71 by the end of its first day on the market, bringing the value of the company to nearly $3 billion. This event demonstrated the massive potential of web-based companies and singlehandedly ignited the dot-com boom.

The following year, Yahoo, with an investment of $2 million from Sequoia, went public on the Nasdaq with its IPO at $13 per share. The company closed the day at $33, a first-day gain of 154 percent, bringing Yahoo's market cap to $848 million. On May 15, 1997, Amazon took its IPO to market with shares priced at $18 and closed the day at over $23, giving the company a market cap of $438 million. For venture capitalists, these successes created a seismic shift in the business landscape as new firms began popping up at an unprecedented rate. Fund sizes began to grow, and nonspecialized VCs overtook those focused on tech at the frontier of innovation.

From 1995 to 2000, VC investment in the US increased from $8 billion to over $100 billion at the height of the dot-com era. VCs were flush with an overabundance of cash like never before, and the sense of FOMO (fear of missing out) was palpable throughout the market between them. Anyone not investing in web-based companies at the time was deemed irrelevant. Firms were racing to invest, wanting to be first in line in what the world now saw as a high-growth industry. Many began writing smaller checks to a high volume of start-ups across a variety of sectors, pouring capital into companies with little more than a quirky name and a domain. This strategy, the "spray and pray" approach, worked well for VCs willing to invest in anyone with user growth metrics (daily active users, accounting rate of return, gross merchandise value) but lacked the technical discipline that had become standard practice before the boom.

The most successful VC players at the time helped define the modern zeitgeist with a focus on speed, high conviction, and founder-friendly GPs. Benchmark Capital, known for its fast decision-making, invested $6.7 million in eBay in 1997, valuing the company at $20 million. Within two years, Benchmark's investment had grown to a valuation of roughly $5 billion. Accel Partners rose to prominence and expanded aggressively through web-based deals with Macromedia, RealNetworks, and a number of other tech start-ups. As the bubble grew, VCs prioritized even shallower metrics over profits in their pool, using page views as primary key performance indicators (KPIs), allowing companies to raise millions without a viable path forward. VCs, driven by hype, FOMO, and visions of fast liquidity, rationalized this recklessness with the belief that the investors at the front of the most lines would eventually come out on top.

By the late '90s, companies like Salesforce began toying with the idea of providing software over the internet as a service. Online infrastructure was still too fragile to support technology like SaaS; most software was still paid for a single time and stored on customer hard drives. While SaaS was developed quietly at the frontier of web innovation, VCs continued to focus on eyeballs and web traffic, not yet anticipating business models with recurring revenue. Few could predict the degree to which SaaS would go on to eat the world of venture.

The Bubble Bursts

The hype-driven mania that fueled the dot-com years led investors to prioritize speed, scale, and exposure over tested, traditional business models. This set dozens of web-based start-ups up to go public on shaky fundamentals. Companies like Pets.com, Webvan, and eToys raised hundreds of millions of dollars without a plan in place to generate and grow profit. VCs that assumed profits would come later for these companies were proven wrong when they didn't. Companies like Pets.com, which posted massive losses following its IPO and went out of business just nine months later, were ridiculed in media headlines, drawing scrutiny from the general public. When the market finally turned in March 2000, the Nasdaq lost 78 percent of its value over thirty months. Hundreds of start-ups were wiped out overnight, leading to immense losses for the investors who had brought them to glory.

The same public that had quickly become enamored by venture's potential soured on the industry as the hype train derailed. Many funds collapsed, and investors began to view VCs as reckless and deceitful. From 2001 to 2003, new VC investments fell by more than 80 percent. A number of key survivors, including Amazon, Google, PayPal, and eBay, were able to thrive throughout the crash by following traditional, time-tested financial strategies. They ran not on hype but on real users, scalable business models, and revenue that showed genuine growth. Google, founded in 1998, intentionally waited until it was profitable to launch its IPO in 2004. These successes proved to a skeptical consumer market that the internet had real business potential but required a patient approach. These years were catastrophic for VC. Dozens of firms shuttered, and deployable capital contracted by 50 percent of its peak from 2001 to 2003. VCs like Benchmark, Sequoia, and Kleiner Perkins that backed dot-com survivors were able to continue raising capital with strong reputations, having demonstrated their ability to exercise discipline amid wild speculation. This led the next generation of firms to focus on capital efficiency, product-market fit, and power-law thinking throughout the 2000s.

Author Tim O'Rielly coined the term *Web 2.0* to describe tech's next frontier. This new bubble revolved around social networks, user-generated content, lightweight apps, and viral social distribution. Companies like Facebook, YouTube, and LinkedIn offered products that improved the more people used them, allowing for growth without massive marketing campaigns. YouTube, with only $11 million in funding from Sequoia, sold to Google in 2006 for $1.65 billion, a win that demonstrated that lost-cost, high-speed returns could be generated once more. Other players included Flickr, Reddit, Digg, and Myspace, all of which contributed to shifting the focus of start-ups from KPIs like page views to metrics covering virality, customer retention, and engagement. Founders began prioritizing business aspects like cost of acquisition (CAC) and lifetime value (LTV), producing what came to be known as lean start-ups.

As optimism began percolating in the markets once more, SaaS hit an inflection point when Salesforce went to IPO in 2004 and raised $110 million for the company. Other early players included Workday, NetSuite, and SuccessFactors. This was great for VCs investing in lean start-ups, as SaaS companies required less upfront capital and were able to measure traction quickly through solid KPIs rather than hype-driven rumors or baseless vibes. Consequently, many investors lost interest in start-ups that ran on specialized technical skills and began relying primarily on spreadsheets to build conviction. The Wild West heat that had powered VC's past cooled into a systematic process.

In response to the 2008 global financial crisis, the Federal Reserve instituted what came to be known as the zero-interest-rate policy (ZIRP) to stimulate economic growth. Over the following sixteen years, ZIRP supercharged VC investing with ultracheap capital, inflated valuations, and influenced start-up behavior. This served as the financial backdrop that set the scene for venture's next big leap.

SaaS Scales Globally as Mega-Rounds Emerge

The 2010s rapidly gained momentum as venture's most explosive decade to date. The high scalability of SaaS and other internet-based businesses made it possible for start-ups to grow faster than ever before. In 2013, Aileen Lee of Cowboy Ventures coined the term *unicorn* to describe start-ups valued at over $1 billion, which became increasingly commonplace as the decade progressed, numbering around five hundred globally by 2019. During this period, VCs laid their eyes on a new frontier: companies that were global, software enabled, and network driven. This time around, they focused primarily on mobile users. Uber, one of the biggest winners, launched in 2009 and blitzscaled internationally with huge capital infusions that allowed the company to raise more than $25 billion before it went public in 2019.

Similar mega-rounds became more common against a backdrop of VC globalization as venture expanded beyond the bounds of Silicon Valley. ByteDance, for instance, founded in 2012 in China, went on to create TikTok, proving Chinese consumer tech could also enjoy meteoric rises on the global venture stage. VCs in India, Latin America, Africa, and Southeast Asia began identifying unicorns of their own, and a wave of emerging-market VCs appeared. Internationally, venture volume rose from $10 billion in 2010 to over $100 billion in 2019.

Despite the proliferation of mega-rounds in VC, GPs of niche funds held their ability to spot and nurture future unicorns well before the herd, capturing outsized returns with small upfront bets. In 2010, one-man micro-VC fund Baseline Ventures wrote a $250,000 seed check to a photo-sharing app idea called Burbn. Burbn soon pivoted into Instagram, and two years later, Facebook acquired the company for $1 billion. Baseline's tiny investment turned into an estimated $80 million payout. Amazingly, Baseline's solo GP, Steve Anderson, identified Instagram's potential and won over founder Kevin Systrom at a time when many larger firms either overlooked the company or feared it was too similar to other bets. This left the field clear for smaller players.

Baseline's deep focus on early-stage consumer web deals gave it conviction where generalists wavered. The result was one of the most iconic venture returns of the decade.[5]

Two additional unicorns were spotted in 2016 when serial entrepreneur-turned-investor Jason Lemkin raised one of the first microfunds dedicated entirely to SaaS, aptly named the SaaStr Fund. With a focused fund size of approximately $70 million initially, Lemkin didn't spray money across industries. Instead, he concentrated on what he knew: enterprise SaaS at seed stage. This discipline paid off. SaaStr Fund backed search-as-a-service start-up Algolia with an initial investment of approximately $12 million[6] in 2014, along with Talkdesk, a cloud call center service. Both thrived and eventually hit unicorn status, returning much of the fund's capital from single exits. Algolia's valuation ballooned to $2.25 billion and Talkdesk grew to around $10 billion as of SaaStr's last funding round in 2021.[7]

How did this tiny fund compete in deals? Lemkin's extensive reputation in SaaS meant founders sought his advice; he often invested *after* helping companies for months via the SaaStr community. While larger funds chased the next trendy fintech or crypto deal, SaaStr Fund stuck to its knitting—small checks in unfashionable B2B software. Over several years, the fund quietly produced top-quartile results. Its performance underscores that a disciplined, thematic strategy can consistently beat a more opportunistic approach. By knowing exactly

---

[5] Power in Numbers (PIN), "The Story of Instagram's First Angel Investor," *Medium*, May 5, 2023, https://powerinnumbers.medium.com/the-story-of-instagrams-first-angel-investor-7cacef62e46b.

[6] John Stewart, "Algolia Advances API-First Software Development; Valuation Soars to $2.25B with $150M Series D Funding," *Algolia*, updated April 16, 2025, https://www.algolia.com/blog/algolia/algolia-funding-news-2021.

[7] Tracxn, "Talkdesk's Funding Rounds," updated July 13, 2025, https://tracxn.com/d/companies/talkdesk/__7vCoun9E-ccRAltQtbHeM0RisJoyuuWnOb5pHgtxENg/funding-and-investors; Talkdesk, "Talkdesk Valuation Triples to More Than $10 Billion, Appoints First Chief Financial Officer," press release, August 12, 2021, https://www.talkdesk.com/news-and-press/press-releases/talkdesk-raises-series-d-funding/.

what it's looking for (and *ignoring* everything else), a focused fund increases its hit rate on the opportunities it does pursue. The SaaStr Fund story is a case study in the power of saying *no* to 99 percent of deals so GPs can say *yes* decisively to the right 1 percent.

By this point in history, VC's superstars had adopted a strategy of marketing themselves similarly to the start-ups they funded. Firms like Andreessen Horowitz (a16z), Sequoia, and Benchmark became full brands with theses and manifestos designed to attract founders who aligned with their visions. Their GPs began connecting with founders and LPs through podcasts, Twitter accounts, blogs, and newsletters, allowing them to position themselves as influencers with established, scalable platforms. Another trend began to emerge as well—VC infrastructure as a service. Platforms like Carta, AngelList, and Crunchbase SaaSified venture with fundraising and management platforms for GPs and start-ups alike. Most VCs focused even more on SaaS metrics rather than on technology itself, deepening the industry's identity crisis as a supposed driver of innovation.

Chapter 4:
The Lost Years

A Look Ahead (The TL;DR)

- Venture's lost years were characterized by frenzied activity, misplaced incentives, and fading differentiation.
- The Federal Reserve reintroduced near-zero interest rates in early 2020 to stabilize markets during the COVID-19 crisis, causing funds to balloon.
- In 2022, policy rates rose from near zero to over 4 percent, bringing an end to the latest ZIRP era.
- Alpha became elusive during the boom years and was generated mainly by smaller funds. Conviction gave way to consensus.

Modern venture has been defined by crisis, correction, and persistent questions about who the industry is ultimately built to serve. For a moment, it looked like 2020 might be the dawn of something new. As the pandemic took hold, many expected markets to crash and capital to retreat. Instead, the opposite happened. Fueled by massive stimulus, near-zero interest rates, and an accelerating shift to digital infrastructure, venture capital entered its most euphoric phase in decades.

The conditions were perfect. Remote work broke down geographic gatekeeping. Zoom fundraising allowed start-ups to raise capital from anywhere, without ever stepping into a partner meeting. A new wave of founders emerged, as did a new class of investors, many of whom had little operational experience but were suddenly flush with capital and conviction. Funds multiplied and valuations soared. More venture capital was deployed in 2021 alone than in the entire five-year stretch from 2005 to 2010.

This was not just a boom. It was the peak of what would later be seen as venture's *lost years*—a period not only of frenzied activity but of misplaced incentives and fading differentiation. The seeds of the correction to come were sown in the very mechanisms that had once made venture capital effective.

How Did Funds Balloon?

The ballooning of VC fund sizes was no accident. It was the inevitable result of prolonged macroeconomic conditions that punished conservatism and rewarded capital risk. When the Federal Reserve reintroduced near-zero interest rates in early 2020 to stabilize markets during the COVID-19 crisis, the second wave of ZIRP (ZIRP 2.0) was set into motion.

This created a flood of cheap capital looking for yield. Institutional LPs, facing limited returns from traditional asset classes, funneled money into private markets at an unprecedented scale. Venture capital, especially, became the asset class of choice.

Mega-funds were the direct beneficiaries of this shift. From 2020 to 2022, over seventy venture firms raised funds in excess of $1 billion. This wasn't just Sequoia or Andreessen Horowitz expanding their reach. It was also growth-stage tourists, crossover hedge funds, and even celebrity-backed firms entering the fray. Venture capital, once a niche, craft-driven asset class, became a magnet for institutional capital-chasing momentum.

At the same time, a huge wave of emerging managers popped up. While some had deep operating backgrounds or differentiated theses, many others were indistinguishable. With capital abundant and LPs eager to back the next breakout, new funds launched with minimal track records, thin theses, and little more than access to the right social circles.

It was a moment that raised a key question: If your firm disappeared tomorrow, would the market notice? No GP who struck gold during VC's early rise had to ask themselves that. But during the ZIRP-era boom, it became increasingly relevant.

Shallow Tech and Herd Behavior

At the same time, the quality of ideas being funded began to flatten. The rise of shallow tech plagued the industry. Teams building SaaS tools with recycled "Uber for X" pitch decks became commonplace. Technical risk was out. Narrative risk was in. This wave of democratized capital lowered the technical and imaginative bar for what counted as innovation.

VCs, now managing bloated funds and pressured to deploy quickly, leaned into safer, faster-moving consensus bets. Conviction gave way to herd behavior. Attend a VC event in San Francisco and you'd hear dozens of GPs repeating the same thesis: "AI for X," "SaaS for Y," or "creator tools with network effects." Everyone sounded the same. No one stood out.

The original spirit of venture (funding the nonobvious, backing weird outliers, taking real technical risk) was being quietly abandoned. True moonshots became rare. Most firms began chasing postconsensus deals, waiting until the market had already signaled a winner before getting involved. By then, prices were sky-high, and upside was limited. But it didn't matter. The momentum would carry the next round, or so they thought.

As capital flooded in, valuation discipline collapsed. Start-ups were rewarded not for building enduring businesses but for attracting follow-on capital. The game shifted from value creation to fundraising momentum. Founders built pitch decks around who was likely to lead the next round, not customer value or revenue growth. It worked—until it didn't.

Our Most Painful Lessons:
The Best Founders Know How to Pivot

We look for founders who are adaptable. Niche is great—but only if the founder knows when to evolve. The best ones use their domain advantage as a launchpad, not a prison. They know when to pivot and when to push through.

The End Arrives Swiftly

By early 2022, inflation had surged to levels not seen in four decades. What began as a temporary supply chain issue quickly became a systemic economic concern. In response, the Federal Reserve ended its zero-interest-rate stance and began one of the steepest rate-hiking cycles in history. Within a single year, policy rates rose from near zero to over 4 percent.

As liquidity tightened, public-market tech multiples collapsed. The IPO window slammed shut. Growth at any price was no longer a viable strategy.

Then came the aftershocks.

Crypto collapsed in mid- to late 2022, culminating in FTX's bankruptcy.

Silicon Valley Bank, long the default financial institution for start-ups and VCs, failed in March 2023.

LPs, already spooked, pulled back from venture as a whole.

Suddenly, the same LPs that had been over-allocating to venture just a year earlier were holding their wallets tightly. And when forced to choose between backing an undifferentiated new manager or re-upping with a brand name, they chose safety. After all, no principal ever got fired for allocating to Sequoia.

But that created its own problem. The lack of new capital for emerging managers and lack of differentiation of the ones that remained only deepened the stagnation. The industry didn't just lose its edge. It lost its imagination.

A Systemic Loss of Alpha

Alpha, the core promise of venture capital, became elusive during the boom years. As firms took on more capital and moved to later stage, they began to behave less like venture investors and more like public market participants. Performance became tied to macroeconomic cycles rather than differentiated insight or conviction. Venture lost its edge. What was once a high-risk, high-reward strategy for backing outliers became a capital-heavy, slow-moving asset class, offering neither meaningful alpha nor useful diversification.

The end of the ZIRP era came swiftly. In the wake of the 2022 crypto collapse and several high-profile bank failures, the decade-long run of easy capital and artificially low interest rates came to a close. As liquidity dried up and market volatility increased, LPs pulled back from risk. Many shifted their capital toward established firms with long track records and recognizable names. No principal at an LP institution would be questioned for allocating to Sequoia, but taking a chance on an unknown emerging manager became harder to justify. In this new environment, venture commitments were shaped less by potential upside and more by institutional defensibility.

The imbalance of capital allocation between established and new funds was staggering. In 2022, the US saw thirty-five venture funds raise over $1 billion each, collectively securing about $91 billion in fresh capital,[8] while emerging managers collectively raised only about

---

[8] Kia Kokalitcheva, "Venture Capital Has Lots of Dry Powder in 2023," *Axios*, January 14, 2023,
https://www.axios.com/2023/01/14/venture-capital-dry-powder-2023.

$15 billion—marking the lowest total since 2015.[9] Unfortunately, this divergence has had negative repercussions for the industry as a whole.

The irony is that alpha was still being generated—just not where the capital was flowing. Small, focused funds with the ability to detect nonconsensus winners at the earliest stages continued to spot companies like Figma, Notion, and Wiz at valuations low enough to deliver 100× returns. But those funds rarely saw institutional commitments during the boom. They were too small. Too weird. Too focused.

In chasing scale, the venture industry lost sight of what made it valuable. The lost years of 2020 to 2022 weren't lost due to a lack of capital. They were lost because conviction gave way to consensus.

---

[9] Pitchbook and National Venture Capital Association, "Venture Monitor: Q4 2024," NVCA, January 2025,
https://nvca.org/wp-content/uploads/2025/01/Q4-2024-PitchBook-NVCA-Venture-Monitor.pdf.

Chapter 5: Rethinking Modern Venture

A Look Ahead (The TL;DR)

- For decades, institutional investors have been pouring money into brand-name mega-funds. This typically garners meager, beta-like returns.[10]
- Most GPs at mega-funds are incentivized by annual fees rather than carried interest. Low hunger can lead to low performance.
- For allocators seeking true alpha, mega-funds are rarely the right vehicle. A grounded, probability-aligned strategy is not only more plausible but also far more likely to deliver real, repeatable outperformance.
- The next great chapter of venture will not be written by GPs of mega-funds. It will be led by GPs of niche funds with the insight, hunger, and discipline to find the outliers first.

Modern venture is plagued by bloated funds and broken incentives. While large, brand-name funds may reflect the historical success of

---

[10] In venture capital, beta-like returns refer to performance that mirrors the average outcome of the overall start-up ecosystem without generating outperformance (alpha). While beta isn't precisely measurable in venture, the term is used to describe funds that follow the market's general trajectory—typically through broad, diversified exposure rather than differentiated insight or access.

venture capital as an asset class, they are no longer structured to prioritize alpha. In fact, many of the industry's largest and most recognizable players have quietly shifted from high-conviction investing to capital aggregation.

This shift has profound implications for both founders and LPs. For institutional investors, there has long been a false comfort in backing brand-name firms. But pedigree does not guarantee performance, and the historical data shows that mega-funds often underperform. Meanwhile, for GPs, the incentives created by fund size and fee structures increasingly reward scale over returns.

To understand why alpha has become so elusive at the top end of the market, we need to take a closer look at the structure and incentives behind the modern mega-fund.

Prestige Isn't a Strategy

For decades, institutional investors have been pouring money into brand-name VC mega-funds, assuming a large fund size and famous GPs would surely yield big returns. In 2009, the Ewing Marion Kauffman Foundation, itself an LP in more than a hundred venture funds, published a blistering report that shattered this myth. Titled "We Have Met the Enemy . . . and He Is Us," the report revealed that Kauffman's VC portfolio vastly underperformed public markets. Astonishingly, not one of Kauffman's large funds (those over $500 million) had returned more than 2× net of fees. The majority of all funds—62 percent—failed to beat even a small-cap stock index. Kauffman's analysis showed that bigger funds often deliver *worse* results, partly due to the fee structure ("2 and 20" management fees incentivize raising capital over generating returns). The foundation candidly admitted that LPs like itself were complicit by blindly committing to oversized funds on GP-friendly terms. But why doesn't pouring more money into established funds work?

Misaligned Incentives: The Principal–Agent Paradox in Mega Venture

To truly understand how mega-funds work, it's essential to consider the incentives that drive their GPs, along with the implications of those incentives. Venture capitalists are paid in two main ways: management fees and carried interest, which work very differently in practice. Management fees, typically charged at 2 percent annually for the duration of the fund, are collected to cover a firm's operational costs, including rent, salaries, and due diligence. These fees are steady, predictable, and unrelated to the fund's performance. Carried interest, by contrast, is a form of profit sharing that typically provides GPs with 20 percent of the fund's total profits after LPs are fully repaid their principal. Few other industries incentivize financial managers to grab a piece of the upside deals to this degree. Carried interest is the element that aligns the interests of all involved parties for the duration of a fund's lifetime.

Originally, it was venture pioneers Arthur Rock and Tommy Davis—cofounders of the firm Davis & Rock—who formally wrote a 20 percent share of capital gains to their GPs in 1961. This arrangement returned around $100 million on a $3 million investment, proving the model worked. It was Tom Perkins in 1972, however, who championed the simple 2 percent management fee and 20 percent carry structure used in venture today. He then pushed this arrangement on Sand Hill Road LPs, cementing it as the de facto standard.

Managers of public assets at hedge or mutual funds charge their fees based on their total assets under management (AUM). This means that if those assets increase, they can then draw fees based on a higher amount of AUM. Venture capital doesn't share this feature, as carried interest can only be based on the original amount investors committed to the fund up front.

The key to understanding the difference between incentives for GPs of mega-funds and small funds becomes obvious when we consider just how much money GPs at mega-funds are regularly making in fees. Mega-fund GPs making millions in annual fees don't need to aim for carried interest, which they receive only at the end of a fund's lifecycle if it happens to be profitable. In the short term, they're incentivized to deploy more capital and raise successively larger funds. Capital-intensive businesses working with mega-funds are pumped with astronomical amounts of funding throughout fundraising rounds. Early founders and angel investors tend to get taken advantage of in this process, as they're unable to keep their ownership in the companies they fund over time. The overabundance of capital drowns out their investments. There's a clear misalignment of interests in these cases, but many GPs at mega-funds aren't driven by alpha-generating outcomes. The management fees they earn on an ongoing fund, not the potential for carried interest down the line, drive them daily when they wake up in the morning. Performance becomes moot.

If a GP's annualized return is similar to the amount they would gain through carried interest, there's very little incentive to perform well. There's no *hunger* for alpha. For a given geography, the average income of a midlevel executive is likely to keep GPs hungry while making five times the average salary, effectively snuffing that motivation out.

Table 3. A Comparison of Income by Fund Size

	Small Fund (e.g., $100M)	Mega-Fund (e.g., $5B)
Total fees (10 years @ ~15%)	$15M	$750M
GPs to split	4	15
Annual fee per GP	$0.38M	$5M
Carry needed to match fees	2.5× fund (3× gross)	1.2× fund (!)

As table 3 shows, for funds above $1 billion, carry is icing, not the cake. This table compares a small fund of $100 million with a mega-fund of $1 billion, illustrating their profits from their funds' final values. The mega-fund GP in this scenario, whose fund earns a 2× return, will receive the same amount in carried interest that they've earned in management fees. The GP of the smaller fund, by contrast, receives a higher return of 3×, enabling them to double the carried interest they'll ultimately receive.

To this day, small venture firms remain incentivized by the potential to receive carried interest through a high-performing fund they manage. The management fees they receive are considerably lower, encouraging them to shoot for high returns at the end of a fund's lifecycle. Smaller funds tend to outperform mega-funds. When this happens, GPs are primed to take home more in carried interest than they have over the course of a decade from all management fees combined.

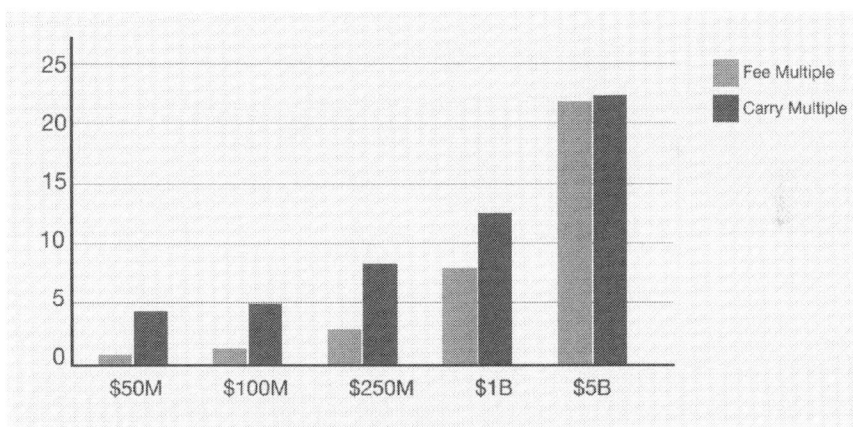

Figure 3. Per-GP Fee and Carry Multiples After Expenses Relative to $300,000 Salary Benchmark, Assuming Top Decile Performance

Smart founders are becoming increasingly aware of this perverse sense of incentives driving mega-fund GPs. In the interest of building their start-ups in more capital-efficient ways, many have pivoted to seeking early capital from smaller funds whose goals align more closely with their own. They may receive less funding initially by taking this route, but they should have better outcomes. Founders who take capital from mega-funds become incentivized to continue doing so, setting them up for greater dilution.

Before the emergence of mega-funds, all VCs were driven by carried interest over management fees. Mega-funds do provide a valuable service to investors looking to deploy large amounts of capital and balance their portfolios with predictable beta returns, but they're no longer truly playing the venture game. Historically, venture capital has been analogous with the high-risk, high-reward pursuit of alpha. LPs seeking alpha should update their approach to VC with a mix of both for better potential outcomes. Investors working with niche VCs are poised to benefit from the renaissance of specialization-focused venture on the horizon.

Why Mega-Funds Miss Alpha

Alpha has long been the reason institutional allocators carve out space for venture capital in their portfolios. Historically, venture was where you went to back the outliers, to take asymmetric risks that public markets couldn't offer. But the rise of mega-funds has distorted that role.

Today, firms deploying billions into late-stage deals increasingly behave like public market proxies. Their performance rises and falls with macroeconomic cycles, not with differentiated insight or early-stage conviction. Venture capital, once synonymous with high risk and high reward, now often produces long-duration assets with characteristics more similar to private equity: low beta, low alpha, and limited upside. This shift fundamentally undermines the role venture is meant to play as a source of outperformance.

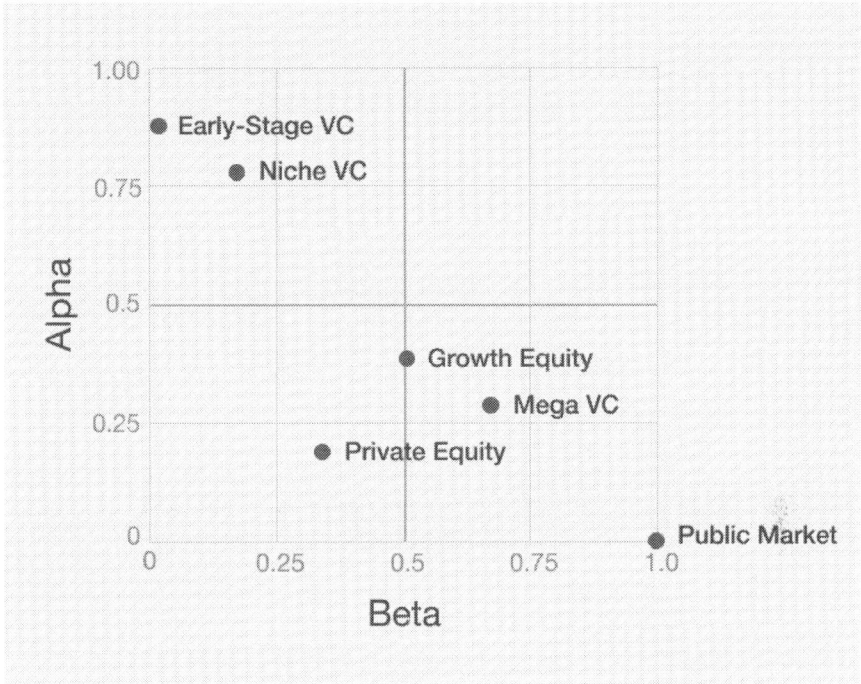

**Figure 4. Return Profiles of VC, PE, and Public Markets
by Strategy**

While GPs at mega-funds aren't aiming for high relative multiples, they are simply trying to generate more value in absolute money. If they have $1 billion of capital to deploy, they could potentially double that money with 2× MOIC. While this wouldn't be a very good MOIC as a whole, this mega-fund would be generating $1 billion in absolute value. This is far more than a small VC could ever make through returning DPI on sub-$100-million fund deals, but still only comes out to returns of 2×. Diversification and investing in larger, later-stage deals may cap both the upside and the downside of larger funds, but their ability to generate raw dollars is unmatched.

For many large institutions, this change is acceptable. Pension funds and sovereign wealth vehicles are often more focused on consistency than on outperformance. When public markets are growing steadily, uncorrelated assets with low volatility (low beta), like today's late-stage venture market, can look appealing. In this model, venture is treated more as portfolio ballast than a driver of generational returns.

But for allocators seeking true alpha, mega-funds are rarely the right vehicle. The most extraordinary outcomes in venture, the 100× returns and transformational winners, almost always trace back to small, focused funds that spotted something early. Funds under $100 million are structurally positioned to enter companies at valuations low enough to deliver exponential upside. They operate with focus, flexibility, and far stronger alignment to founders and LPs.

Mega-funds still have a role. For investors seeking stable, midrange returns in the 5 to 15 percent range, they function more like private equity. But for those chasing outliers, innovation, and real alpha, the frontier lies elsewhere.

How to Spot an Implausible Fund

Before backing any fund, smart LPs ask a simple question: *Can this strategy actually work?* One simple way to test whether a fund's strategy can actually live up to its promises comes from Josh Kopelman, GP at First Round Capital. He created the Venture Arrogance Score (VAS) as a quick litmus test for plausibility.

The Venture Arrogance Score

A simple formula to evaluate a VC firm's business model

$$\frac{\text{Fund Size} \div \text{Exit Ownership} \times \text{Target Multiple}}{\text{Investment Window} \times \text{Annual Startup Value Creation}}$$

The idea is straightforward: Compare a fund's target ownership, fund size, and return expectations against the total annual exit value of the US venture market. Why does this matter? Because no US fund family—not even Sequoia, a16z, or Insight—has repeatedly captured more than 10 percent of national venture exit value. Venture is, by nature, a fragmented market. So when a new vehicle's strategy implies it needs to capture 15–30 percent of all exits to hit its targets, it's effectively betting it can outperform the entire field of top-tier firms combined. That level of dominance is unprecedented. It's the venture equivalent of a mutual fund claiming it will triple the S&P 500. LPs should see that not as ambition but as a red flag.

Sophisticated LPs increasingly use the VAS as a first filter. If the number looks superhuman, the rest of the deck probably is too. Many then follow up with a binomial or Monte Carlo model to evaluate how ownership, reserve strategy, and expected loss rates affect the actual probability of success.

To ground the concept of the VAS in reality, I ran the numbers on three common scenarios: a $7 billion mega-fund, a typical $100 million seed fund, and our own $25 million fund at Omni Ventures. Using conservative assumptions on ownership, target returns, and investment pacing, the VAS reveals just how much of the total US venture exit market each strategy must capture to succeed.

Table 4. Venture Arrogance Score Calculated for Three Fund Sizes

Scenario	Fund Size	Exit Ownership (%)	Target Multiple (×)	Invest Window (Years)	Annual Exit Value (US)	VAS (%)
Mega-fund	$7B	10	2.5	3	$200B	29
Typical seed fund	$100M	10	3	3	$200B	0.5
Omni Ventures	$25M	7	3	4	$700M	0.38

Note: The assumptions made about Omni Ventures are as follows: $25 million final close, 12 percent ownership, and 3× gross.

According to the VAS score, the mega-fund must capture 29 percent of the total value of US venture exits, which is very aggressive. The typical seed fund needs one or two fund returners, which is historically feasible. Omni Ventures requires only 0.38 percent of the total value of US venture exits, which is comfortably below historical top-quartile capture rates.

The results of the analysis speak for themselves. Mega-funds need to consistently dominate nearly a third of all national venture outcomes, a bet that no firm in history has sustainably pulled off. By contrast, typical seed funds and especially focused niche funds operate well within historically achievable ranges. With a VAS under 0.5 percent, a niche fund doesn't require heroics. It just requires doing their job well. For LPs, this kind of grounded, probability-aligned strategy is not only more plausible, but also far more likely to deliver real, repeatable outperformance.

Niche Is the Next Frontier

Venture capital has drifted off course. In the rush to grow, many firms have sacrificed focus for scale and conviction for consensus. Mega-funds have ballooned in size, chasing larger rounds and optimizing for assets under management rather than differentiated insight. At the same time, a wave of indistinguishable early-stage funds has entered the market, crowding into the same deals without clear theses or ownership discipline. What was once a field driven by bold ideas and sharp judgment now often feels like a volume game.

Mega-funds in particular attempt to scale across all stages and sectors, offering the same capital and support to every founder. In trying to be everything to everyone, they dilute their edge. These firms provide access, but not insight. And in a market where capital is no longer scarce, generic money has become just another commodity.

This shift is pushing a new movement. A growing class of LPs is seeking something sharper and turning toward specialist funds. These smaller, focused managers have deep domain knowledge and the conviction to act early. These funds do not rely on consensus to justify a deal. They do not need to chase what is already obvious. Instead, they earn their returns through discipline and insight, spotting breakout potential long before others understand what they are seeing.

In many ways, this is a return to the origins of venture capital. The industry was built by experts backing frontier technologies that required specialized knowledge to even recognize. The biggest wins rarely came from safe bets. They came from conviction calls: unpopular at the time, obvious only in hindsight.

The next great chapter of venture will not be written by those managing billions. It will be led by sharp, focused funds with the insight, hunger, and discipline to find the outliers first. Niche is not just a strategy. It is where the future of alpha now lives.

Chapter 6:
The Rise of the Niche VC

A Look Ahead (The TL;DR)

- Small, sector-savvy funds punch far above their weight. LPs seeking real alpha increasingly allocate to firms that see around corners in their chosen domain.
- Niche VCs consistently outperform thanks to a combination of faster decision-making, stronger founder alignment, and clearer risk exposure.
- Each niche firm isolates a blind spot, builds earned secrets within it, and scales only to the level the niche can actually absorb.

As we discussed earlier, the power of niche lies in the perspective it creates. Specialists stand on a different hilltop; they see the next ridge line while the crowd below is still chasing the last rumor. When generalist firms swarm each new fad—crypto yesterday, AI infra today—niche VCs quietly excavate opportunities that nobody else realizes are gold yet.

Back in 2015, the term *pre-seed* was barely on anyone's lips. Yet in a converted Brooklyn warehouse, Notation Capital raised a tiny $8 million debut fund dedicated to writing the very first $250,000–$500,000 checks to New York founders. Six months later, on the opposite coast, Afore Capital closed its $47 million Fund I with the same radical promise for Bay Area start-ups: "We're your first institutional partner—before the series seed even has a term sheet."

Both funds were born from a simple insight: The earliest capital market had gone missing. Seed funds had drifted up market, angels lacked capacity, and founders were stranded in a yawning funding gap. By laser-focusing on a stage when cap tables are clean and valuations rational, Notation and Afore earned proprietary deal flow and board-level influence long before bigger firms woke up. Their bets—Notation on Alloy, Chainalysis, and Bounce; Afore on Flatfile, Modern Treasury, and Pie Insurance—have since generated greater than 3× gross TVPIs on their first vintages and validated that niche pre-seed is now a distinct and established asset class of its own.

Founders never forget their first check. When we launched Afore back in 2016, "pre-seed" sounded like a punchline—founders literally asked, "WTF is pre-seed?" Once we framed it as the first institutional check plus hands-on help, the lightbulbs went on. A few years later everyone was branding a "pre-seed" strategy. Venture is awash in capital now, but conviction is still scarce: The best founders remember forever who believed first, and they choose partners they meet through trusted, warm introductions—not anonymous marketplaces. With AI resetting the playing field, the edge still belongs to the fund willing to bet before the crowd and then roll up its sleeves.

—Gaurav Jain, Cofounder and GP, Afore Capital

Modern niche VCs like Notation and Afore demonstrate how focus outperforms size. While mega-funds spray capital across every trend and stage, specialists build conviction flywheels: deeper diligence → better hit rate → stronger reputation → even better deal access. That cycle is why small, sector-savvy funds punch far above their weight and why LPs seeking real alpha increasingly allocate to the firms that see around corners in their chosen domain.

**Our Most Painful Lessons:
Bet on Founders, Not Just Companies**

Your first check isn't just for this start-up—it's for the next one too. If the founder fails but remembers you were there early, you're first in line next time. Great founders keep building.

Tactical Edge: Why Niche Works

The performance advantage of niche funds isn't just theoretical—it is deeply tactical. Specialized funds consistently win deals, drive outcomes, and return capital faster because their focus gives them a real operational edge. Niche GPs don't need weeks to ramp up on a sector, because they are already immersed in it. That allows them to move quickly and issue term sheets with genuine conviction while generalists are still conducting reference calls. Their deep domain expertise enables sharper pattern recognition, helping them spot early signals that matter and that filter out noise. This leads to better pick rates and fewer false positives.

But perhaps the most important differentiator is how founders perceive them. Founders are increasingly wary of capital without context. They don't just want money—they want partners who understand their market, anticipate their challenges, and can offer meaningful help when it matters most. Niche GPs speak the same technical language, have relevant operator experience, and bring networks that are already aligned to the start-up's needs. Whether it is navigating regulations, refining a go-to-market plan, or recruiting specialized talent, niche VCs provide support that is targeted and hands-on. Founders see them as an extension of their team, not just another name on the cap table.

Our Most Painful Lessons:
Be the Investor They Make Room For

In early-stage venture, the best deals aren't won on price—they're won on trust. We've consistently gotten into sub-$10-million rounds not by negotiating founders down but by being the investor they genuinely want on the cap table. When you show up early, offer sharp insights, and build a reputation for being responsive and founder aligned, you become part of the company's success story, not just a line on the cap table. Founders are often willing to flex on terms when they believe you'll meaningfully increase their odds of winning. The goal isn't to squeeze value out of a deal. It's to be so valuable that they make space for you.

For LPs, the benefits are just as clear. Sector-focused portfolios provide more coherent and intelligible exposure, making it easier to underwrite risk and avoid style drift. Instead of betting on a few winners scattered across hundreds of spray-and-pray generalist investments, LPs gain clarity into a defined and repeatable slice of alpha. The combination of faster decision-making, stronger founder alignment, and clearer risk exposure is why niche VCs consistently outperform their weight class.

Niche Success Stories

Specialist franchises continue to beat expectations: Each firm isolates a blind spot, builds earned secrets within it, and scales only to the level the niche can actually absorb—never to the point of diluting expertise.

Lux Capital

Lux Capital is a prime example of a niche fund that carved out alpha through deep specialization and contrarian conviction. Originally focused on the intersection of science and frontier technology, Lux built a reputation for backing deeply technical, hard-to-understand start-ups well before they became consensus bets. Rather than chasing hot deals, Lux cultivated long-standing relationships with academic researchers, national labs, and technical founders—giving them proprietary access to transformational ideas years before others took notice.

One of Lux's standout investments was in Zoox, the autonomous vehicle start-up acquired by Amazon for over $1.2 billion. Lux backed Zoox early, not because it fit a trend but because they believed in the technical edge of the founding team and the long-term disruptive potential of a vertically integrated approach to self-driving cars. Similarly, Lux was early in Planet Labs, the satellite imaging company, and Auris Health, which was acquired by Johnson & Johnson for $3.4 billion. These wins came not from spray-and-pray investing but from a rigorous, science-first approach that led to category-defining outcomes. Lux's edge lies in its ability to understand and underwrite technological risk where others see opacity—and that has translated to meaningful outperformance.

As major VC firms scale into multi-stage, multi-strategy, multi-geo platforms, a new class of focused, exclusively early-stage VC firms is poised to emerge. I believe founders want VCs who can provide capital AND would take active roles (such as board seats) in helping hire, develop GTM [go-to-market] strategies, navigate regulatory and related complexities, and help in capital formation. Venture as a craft will return as other VCs industrialize themselves into heat-seeking missiles.

—Bilal Zuberi, Founding Partner, Red Glass VC (previously scaled Lux Capital from $245 million Fund III to $1.15 billion Fund VIII)

Susa Ventures—Company-Building Infrastructure

Susa Ventures has quietly become one of the most consistent seed investors in Silicon Valley by leaning into infrastructure, not hype. The firm specializes in backing technical founders building foundational tools and platforms—companies like Flexport, Robinhood, Rippling, and Loom. What sets Susa apart isn't just deal access but hands-on company building. The firm helps lay the groundwork for scaling: product hiring, financial modeling, customer development. Their focused operational support gives them a strong reputation among engineers and repeat founders, many of whom choose Susa even when they could raise from flashier names. In a world chasing vertical hype cycles, Susa wins by focusing horizontally on how great companies are built.

Part III:
A How-To Guide for Niche

Chapter 7:
How to Pick a High-Performing Niche VC

A Look Ahead (The TL;DR)

- Just as founders make up 90 percent of a start-up's value, GPs are the crux of emerging venture firms that soar.
- Great GPs are intrinsically motivated and well-rounded and operate according to an ethos of reciprocity.
- Proprietary deal flow enables GPs with strong networks to break through the noise and grow stable funds.
- Savvy GPs stay aware of changing market conditions and balance the need for DPI with IRR.
- While fundraising, GPs must connect with committed anchor investors to generate momentum. They can do this by offering anchors certain preferential terms.

While software VCs chased consumer apps, Palo Alto–based Eclipse went after supply chain, manufacturing, and energy hardware. Its 2019 vintage fund hit a 1.74× TVPI by mid-2022 with 84 percent of capital deployed, an impressive mark for such capital-intensive arenas, and the franchise has since scaled to $4 billion AUM. Eclipse's edge came from operator GPs who could underwrite factory-floor risk that scared traditional software investors.

When cash is flush in venture, as it was during the ZIRP era, scores of new funds pop up, led by emerging GPs without a thoughtful strategy in place. When money comes easily, those without grit or discipline may find themselves straying from choices that prioritize the best interests of their allocators. As with all venture booms, the bubble eventually bursts, resulting in extinction-level crashes likely to wipe out fledgling firms. Those who manage not only to survive such periods but to gain traction for their funds in spite of them are more likely to strike gold in the long run. A savvy, dynamic GP is often the single most important factor contributing to funds that soar rather than sink.

Our Most Painful Lessons:
You're the Product

Venture is a trust game. LPs are backing *you*. They want to believe you'll be a good steward of their capital and a force multiplier for founders. That means selling not just your thesis but also your judgment, grit, and vision. You're not raising money for a fund— you're raising belief in yourself.

If founders are 90 percent of a start-up's value, the same can be said for GPs of emerging venture firms. While we technically invest in business entities, in truth we are betting largely on the people who run them. Founders and GPs both become successful by convincing people they have an edge no one else understands. To grow their companies and careers, both must constantly hustle from a place of relentless grit and be ready to pivot on a dime according to market needs. In the constantly evolving world of venture, it's essential for GPs to be dynamic, keeping a finger on the pulse of the industry as it fluctuates. Their thesis is likely to shift over time, becoming broader or more niche, but certain qualities of a great emerging GP, like a great emerging founder, must remain static if they're to create value for investors.

The Traits of an Effective GP

Above all else, the most powerful characteristic of a thriving GP is intrinsic motivation. To keep an ear to the ground of the new frontier and make connections with founders primed to generate alpha, a GP must be absolutely relentless in their pursuits. It takes an insane amount of grit, passion, and determination to build a prosperous fund. For these GPs, venture isn't a hobby but an all-consuming obsession. These are people who likely sacrificed the salary of a comfortable day job to solve the biggest pain points the world faces. Any LP seeking a promising fund will benefit from a conversation with its GPs to determine their level of devotion to their firm.

Exceptional GPs are also well rounded, using a strategic mix of both business and sector knowledge and skills. Someone who was once a great engineer won't necessarily make a great investor. Grit won't suffice if you don't know how to run a business or don't have the ability to crunch numbers correctly. GPs are, in essence, founders selling a unique financial product who must be able to conceive a path to revenue over the course of ten years. This requires an investment strategy that differentiates the fund's thesis, provides an edge over the market, and ultimately gets profit back into the hands of investors at the multiple they're looking for. Allocators evaluating GPs should look into whether those managers have previously served in roles beyond that of engineer.

When I cofounded Omni Ventures, I was a technically inclined engineer but lacked experience in financial management. We started out incubating start-ups by helping them develop their products from an engineering perspective. That experience, while valuable, didn't help me become a better investor, because I was still looking at start-ups through an engineer's lens. Growing as an investor meant no longer resting on my laurels. I had to go outside my comfort zone and find the answers to questions that went beyond the quality of start-up products. How would we get our first customer? What pain point were we solving for our customer base? Was our customer base large enough? What could we do to communicate clearly with our founders and investors? How could we get our first million in revenue? These are questions most engineers won't be able to answer off the bat, but they're all incredibly important for LPs evaluating GPs of emerging funds.

Great fund managers also operate within an ethos of reciprocity. These GPs are incentivized to engage their allocators through win-win arrangements that benefit everyone involved. Relationship building is an innate part of human nature, but many venture capitalists in the US skip this crucial step, hungry to dive in and deliver their pitch immediately. This transactional approach signals a lack of focus on the allocator and their goals. It's also an inefficient strategy for cementing a ten-year relationship. A GP can't properly earn trust without first establishing the specific ways they can help an LP. There are no wins to be had if their interests don't align.

Our Most Painful Lessons:
Raising a Fund Is a Trust Exercise

Asking someone for money is the ultimate act of trust. If you haven't built that trust first—through time, consistency, and insight—don't be surprised when the answer is no. Trust isn't built in the pitch deck. It's built over the course of weeks, months, and years.

GPs who prioritize reciprocity in relationships initiate conversations with curiosity about the other party. They resist the urge to open their pitch deck during the first meeting, opting instead to listen to the other person's introduction and ask strategic questions. What's the allocator's background? What interests them about the firm? What are their goals over the next ten years, and how do they envision the firm helping them get there? At Omni, we've had conversations with investors with all kinds of different goals:

- Family offices wanting to learn about technology in venture so they could apply those insights to their investments on public markets

- Institutional LPs wanting introductions to other LPs
- GPs wanting introductions to founders and deal flow
- Corporate venture capital firms (CVCs) wanting to collaborate on research reports
- Family offices wanting advice on how to invest in other asset classes
- International LPs wanting immigration advice due to our large international base
- International LPs wanting advice on how to establish a brand presence in the US to expand their business

While none of those conversations had to do with investing in our fund, we were able to help those people because they realized Omni would make a great addition to their team in some capacity.

Proprietary Deal Flow Is King

Allocators looking to invest in niche-VC funds, which are often smaller and have an early-stage focus, confront a fundamental consideration while evaluating GPs: How do those GPs get their deals with start-ups? This is especially important to think about with newer firms that haven't yet established themselves as a brand. How does a relatively unknown entity make connections in the saturated world of VC? Founders can't find them online using services like PitchBook or Crunchbase or Carta's databases yet. How do they break through the noise? The answer is their network.

A number of successful GPs of specialized VCs have built their firms alongside personal platforms. Chris Yeh of Blitzscaling Ventures, coauthor of *Blitzscaling: The Lightning-Fast Path to Building Massively Valuable Companies*, used his book as a way to land speaking gigs on stages all over the world as he grew his fund. Harry Stebbings of 20VC, who started angel investing in his early twenties, made his splash in the venture scene with *The Twenty Minute VC* podcast. By hosting interviews with expert investors, Harry became a thought leader in the venture space, opening the floodgates for a stream of proprietary deal flow.

Our Most Painful Lessons:
Start Early

I wish I had started angel investing and talking with start-ups in my late twenties like Harry Stebbings did. Just attending events or writing small angel checks would have vastly accelerated my network-building efforts.

Start early.

DPI Strategy

Before the ZIRP era went into effect, VC funds focused on managing long-term capital of ten to fifteen years on average, seeing huge returns upon IPO or acquisition. During the SaaS era while ZIRP was in effect, we saw expectations to liquidity shorten as people became enamored with shallow tech. SaaS consumed the world of venture, proliferating at a tremendous rate. When ZIRP ended and the crash of '22 hit, investors who over-allocated to VC saw their chances at liquidity evaporate. This has caused those LPs to over-index on liquidity since then, prioritizing DPI over IRR. Throughout the ZIRP era, LPs had seen impressive IRR markups on paper that ended up amounting to Monopoly money. They saw so little liquidity returned previously that they lost confidence in the ability of VCs to return capital. Today's VCs face a market demand for earlier DPI. At Omni, we've developed a strategy to create earlier returns to our fund as our portfolio companies exit.

Fund managers have several tools to deliver early DPI to their LPs. One approach is to build a portfolio that includes a mix of start-ups, including those with shorter time-to-liquidity profiles. Another is to leverage secondaries, selling stakes in promising companies ahead of a public exit. M&A activity can also provide quicker returns, particularly in markets where acquisitions are more common than IPOs. Whichever path is used, GPs must stay attuned to market conditions and strike the right balance between generating DPI and preserving IRR. In today's market, demonstrating early DPI not only boosts fund performance but also signals to LPs that the manager is proactive and aligned in returning capital. Ensure the managers you're backing keep an eye on the macro as they're building their fund strategy.

The Ability to Fundraise

Fundraising takes time, especially for GPs working with a blind pool. Different LPs will have different levels of risk tolerance and therefore want to invest either earlier or later. We GPs can control a lot in venture, but we can't control timing. We may meet an LP who's a perfect fit for our fund but miss the chance to work with them for reasons outside our control. They may not have liquidity to enter the fund when we want them to due to unforeseen circumstances. There's no point in losing sleep over missed deals due to misalignments in timing. This is a long game, and allocators who hold back during earlier rounds may end up circling back around later. By staying in touch and keeping potential LPs up to date on the momentum of the fund, GPs can keep the door open to future collaboration.

Our Most Painful Lessons:
Relationships Compound

In early-stage venture, every interaction is a long game. The person you meet at a random event could send you a killer deal two years later. A founder who passes today could become your next scout. There's always latent value in the ecosystem. Don't burn bridges. Don't close doors.

Let's talk now about how GPs can navigate talks with their anchor LP. In a typical fundraising process, GPs do their best to get a minimal commitment from the people in their close network. Ideally, they'll have been networking with LPs for years at conferences and other events. Once they've established this initial base, GPs must then choose whether to stick to it or create a more institutionalized fund size. I prefer to start doing angel investing and implementing an informal track record fund in the form of individual deals called syndicates. This can be done by finding a target asset locally and bringing friends to invest in that one deal. This is a great way to build trust with your network, though you may find you exhaust your base quickly by doing this regularly.

At a certain point, GPs need to start bringing in bigger investors, but I've found many writers of larger checks don't want to come into funds at early closings. Anchor investors writing checks greater than 10 percent of a fund target size are absolutely critical for getting the ball rolling by generating momentum. For small funds, anchor investments of $10 million to $100 million are the sweet spot for generating alpha. Anchor investors can come in the form of large family offices, sector giants, or funds of funds (FOFs). Once a VC has its anchor, you are derisked in the eyes of other potential LPs, allowing them to feel more comfortable investing in the fund. The anchors themselves, once they've got skin in the game, should also be incentivized to introduce the GPs to members of their own networks. This symbiotic relationship is synergistic for all involved.

Our Most Painful Lessons:
The Power of an Anchor

If you want to build a truly institutional fund, you need an anchor. We were ambitious from day one, courting LPs who could unlock the right kind of momentum—but no one wanted to move first. The moment we landed Allocator One as our first anchor LP, everything shifted. It gave others permission to get off the sidelines. Every anchor has a reason to anchor, and you need to understand what angle you can offer: coinvestment rights, a GP stake, a fee discount, or strategic value. Figure that out early.

As we saw in chapter 1, the VCs best primed to find alpha are those that hit all three prongs of the niche flywheel: mastery, focus, and network. They're managed by GPs with big networks and relentless drive who are well rounded, resilient, and focused on building relationships based on reciprocity.

After enough time investing as LPs, some allocators decide to do more direct allocations and take on a role similar to that of a GP. We'll talk in the next chapter about how GPs can handle this leap strategically.

The Only Two Questions That Matter for LPs

When evaluating a venture fund, sophisticated LPs often boil it down to just two questions:

1. Do the GPs have real skin in the game?
2. Are they still hungry?

To assess skin in the game, LPs should ask, *What percentage of your liquid net worth is committed to the fund?* This isn't about absolute dollars. It's about relative sacrifice. A $500,000 GP commitment means very different things depending on whether the GP is worth $1 million or $50 million. The greater the personal stake, the more incentive a GP has to protect and grow the fund's capital like it's their own, because it is.

To gauge hunger, ask the GP, *How much of your potential earnings come from carry versus management fees?* Many managers, particularly those at large or legacy firms, can coast on fees alone. But real outperformance comes from GPs who need to earn their upside. If a GP is positioned to make more from carried interest than from salary, you can bet they're showing up every day to win.

Our Most Painful Lessons:
The Right People Work for the Upside

This isn't a job for people chasing salary or upfront cash. Venture rewards long-term belief, not short-term paychecks. If your early team members aren't energized by equity or carry, they're probably not in it for the right reasons. You need partners who see your upside as theirs and are ready to earn it. Real commitment doesn't come with a paycheck. It comes with shared ownership.

In the end, venture is a long, volatile, emotionally taxing game. The GPs who endure and outperform are the ones with something to prove, something to lose, and everything to gain.

Chapter 8:
Designing Your Own Niche Strategy

A Look Ahead (The TL;DR)

- After years of investing, many LPs start thinking about launching their own fund. If you're dreaming of becoming a GP, there are strategies you can follow to grow a firm that generates alpha.
- To thrive in today's saturated VC market, a firm must assert mastery over a distinctive niche with a unique hilltop advantage.
- Staying disciplined as a GP is incredibly hard but necessary for differentiation and alpha generation.
- A strong network is the key to developing abundant proprietary deal flow. Never disregard a potential LP when it's possible to leave the door open.
- Generalist funds can be classified as niche as long as they adhere to the three-prong flywheel introduced in chapter 1: mastery, focus, and network.

Omni Ventures, the firm I cofounded in 2021 with Sabrina Paseman, specializes in manufacturing-tech start-ups in digital engineering tools, robotic software, factory automation software, and supply chain tech. With all the post-ZIRP-related hardships that existed in 2022, we were essentially starting from negative one rather than zero. We began with a team of industry expert engineers who had been steeped in the industry for over a decade each, but that was not enough. What I quickly discovered was that—similar to how most engineers do not make good, business-savvy founders—most engineers also do not make good VC investors. We refined and downscaled the team to just a few with extreme grit and passion for investing who were willing to put up with the extreme opportunity costs and who were driven to generate outsized returns. Manufacturing is a legacy industry developed before the arrival of the digital age, making tech integration a serious challenge. The frontier of innovation we tackled revolves around the question of how we can digitize aspects of real-world supply chains. How can we capture large amounts of noisy, unstructured data pulled from factory floors and industry at large? How can we use modern AI-powered software to make sense of that data and analyze key insights that were previously lost in the noise? We now find founders working on answers to these questions as they push the boundaries of integrating technology into real-world infrastructure.

When I made the decision to cofound Omni Ventures with Sabrina, I knew that would mean walking away from a million-dollar annual big-tech compensation package. Ever since I'd started angel investing while working at Apple years earlier, I'd been energized in the presence of start-up founders. They were the most passionate people in the world, and being able to work in their orbit was intoxicating. I started imagining myself as a GP, managing my own fund with the privilege of being able to pick my colleagues. As an engineer, I had been Apple's innovation expert when problems arose that our team couldn't solve, resulting in over fifty patents in different areas. I got good at learning about novel skills quickly and effectively, whether it was forming glass into a certain 3-D shape or polishing aluminum into a mirror. I loved the experience, and the thrill of working at the frontiers of innovation followed me. I left Apple to work for start-up ARRIS Composites, which I helped scale during the ZIRP era. We took the company from seed stage to a $290 million valuation in less than two years. In light of that success and the job offer from Meta, I woke up one morning thinking, *It's time. If I keep putting this off, I'll never start my own fund.* After a few months of deep research into manufacturing-tech start-ups, I realized there was a gap in the market I could fill. A need for an investor like me. Soon after, Omni Ventures was born.

After years of investing in VCs as LPs, many investors start thinking about launching their own fund. If you're dreaming of becoming a GP, there are strategies you can follow to grow your firm in ways that are both sustainable and primed to generate alpha.

Mastery: Specialization Choices

To stand out among the crowd in this saturated market, VCs must now pick a one-of-a-kind differentiator that sets them apart from every other competitor. With every new wave of the tech du jour—SaaS, crypto, and now AI—hundreds of funds rebrand themselves to serve the new wave. GPs with genuine expertise in a specific area who know how to market their skills are more easily identifiable by founders who are a match. This is often the secret recipe to a strong brand and proprietary deal flow. The path to alpha lies in being as visible and recognizable as possible to those people in a murky sea of options.

For a GP, the solution to commoditization in venture is to focus on your niche's edge and build a strong and unique brand around this, and in so doing, become as memorable as possible in the eyes of founders, peer VCs, and LPs. How can you grab and keep the attention of the founders you want to work with? How can you position and market yourself in ways that make your firm easy to find for those who are looking? We'll talk about how GPs can effectively platform themselves after this chapter. For now, the main point to keep in mind is that niche VCs that actually generate alpha have an unfair advantage in some way, and that advantage naturally extends to their LP investors. Notation Capital, for instance, became one of the first VCs to institutionalize angel investing by coning the term *pre-seed* venture, but there are countless other ways to specialize. Let's look at some of the most common ways today's niche-VC firms are differentiating themselves.

To thrive in a saturated market, a VC must own a distinctive niche— an unfair advantage both founders and LPs can articulate in a single sentence. In table 5, I map the most common (and a few imaginative) differentiation vectors. Remember Han's finding: Industry and stage focus drive the strongest performance uplift, while geography matters less unless it conveys proprietary access.

Table 5. Differentiation Vectors

Vector	What It Means	Why It Works	Examples
Sector specialists	One industry—fintech, robotics, climate tech, etc.	Deep network, pattern recognition, preferred-buyer exits	**Omni Ventures** (manufacturing tech), **Lux Capital** (frontier tech)
Location focused	One region, country, or city	Ground-game sourcing, civic incentives, talent density	**The House Fund** (UC Berkeley ecosystem), **Kaya VC** (Central & Eastern Europe)
Stage specific	Pre-seed, seed, series A, or growth only	Valuation discipline, playbook reuse, board chemistry	**Notation Capital** (pre-seed), **Afore Capital** (pre-seed)
Thesis driven	Invest according to a worldview (e.g., climate resilience)	Clear brand, attracts missionary founders	**Lowercarbon Capital** (climate), **HCVC** (hardware), **Third Sphere** (climate resilience)

Table 5. Differentiation Vectors *(cont.)*

Impact/ environmental, social, and governance (ESG)	Dual mandate of profit + measurable good	Access to mission-driven talent & capital	**Slauson & Co., Collab Capital, Blue Horizon**
Founder demographic	Back under-represented builders	Authentic community + overlooked alpha	**Backstage Capital** (under-represented founders), **Chingona Ventures** (Latinx founders)

Focus: Discipline Is Key for Differentiation

It's essential for emerging GPs to learn that greed, hype, and FOMO drive 90 percent of the VC industry. When investors see certain funds blasting off like a rocket, they see others making money and follow the herd—even when the frenzy is based solely on hype. To make great deals, you must always take a measured approach and stick to your investment plan, resisting the urge to follow lemmings off a cliff. Avoid getting sucked into fears of missing out on the current trend, and be meticulous about moving forward consciously. At Omni, our strategy requires us to invest in start-ups at the pre-seed stage because our goal is to exit early through M&As. Straying from that model would be a betrayal to our customers, something only a bad investor would do.

Never let the noise and hype distract you as an investor, even when it's coming from your GP peers. Instead, put your blinders on and focus on your fund model. If you choose to have a specialized thesis (and I recommend you do), this will give you the advantage you're looking for. By keeping your eye on the frontier of your specific sector, you'll be able to spot "super frothy" deals before any other firm. The under-the-radar opportunities other investors haven't spotted yet. The best way to stay focused is to pick a lane and stick to your niche, if you have one.

Staying disciplined as a GP is incredibly hard, but it is necessary if you're to differentiate your firm and generate alpha for your investors. It's an open secret that a huge percentage of VCs that emerged in recent years haven't taken the financial aspects of investing seriously enough. These Silicon Valley cowboys rush in to invest during market booms and convince themselves they have no time to waste on crunching numbers. If they sat down to do the math, they would begin to understand why following the herd won't bring them alpha in the long run. They lose sight of the fiduciary duty obliging them to return capital to the LPs paying their management fees and end up with lukewarm returns.

Being disciplined as a GP starts with deciding what you want to be disciplined about. In the case of an early-stage niche VC like Omni, a GP has to understand what types of companies they can and can't invest in. Investing in start-ups outside of our sector thesis could throw our whole portfolio off track. Say, for instance, your firm invests in deep tech, but a fintech opportunity arises that seems too good to pass up. As an investment manager, you have to know how including that start-up in your portfolio could affect your fund returns. I'm always astounded by the number of emerging GPs who fundraise without a detailed fund simulation that helps them anticipate how their returns will perform. Understanding your business at a deeper level than most GPs is likely to drive alpha. This practice of financial discipline is, in itself, a strong differentiator that will help your fund stand out.

How to Pick Your Fund Size

A bigger fund isn't a sin—running past your alpha capacity is. For a GP, choosing a fund's size goes beyond calculating how much you want to earn in management fees. (Wouldn't it be nice if it were that easy?) Here are some other points to consider:

1. How much capital can your niche absorb before the market loses its edge?
2. What company valuation can you actually get access to?
3. Look back at the deals you would have liked to do over the past three years. At what valuation did you find out about them? What size check or ownership amount would you need to have written to get a good return on your fund?
4. How much ownership can you get?

Answering these questions can go a long way in helping you choose a fund size that stays within your capacity to generate alpha. There are many ways to scale a niche fund, but you must always be careful not to scale beyond what your niche can absorb. Deploying billions of dollars into an area without proper deal flow can lead you to become undisciplined, raising your chances of meager returns.

Ensuring Your Thesis Stays Disciplined

The following tips can help you stay focused while scaling a niche fund.

1. **Thesis drift:** If a start-up is outside of the fund's vertical thesis, participation requires unanimous GP consent and limited partner advisory committee approval.
2. **Ownership drift:** If the deal is outside of the fund's target valuation, move it out of the main fund and offer it as a special purpose vehicle (SPV) to your LPs.
3. **Support drift:** Only back companies in which your fund's expertise or network can materially move the needle. If you can't add real value post-check, you're not just drifting—you're diluting your brand.

What Drives a Fund's Focus Depth?

It's commonly assumed that a larger fund size and a great track record are good ways to underwrite VCs. Don't fall into this trap. While the ability to raise a large fund does speak to the GP's ability to fundraise and connect to a strong LP network, it does not speak to whether the GP will have a good strategy. Instead, consider the following:

1. The GP's risk aversion
2. The level of risk involved in the specialty area
3. The specialization of the GP team

Resist the urge to invest solely in smart people with knowledge of a specific technical industry. Make sure they also understand how their knowledge can be plugged into the larger zeitgeist and how to build successful businesses to fill those gaps.

Network: Building a Team That Punches Above Its Class

Building a niche fund requires VCs to stay small, under $100 million in most cases, meaning GPs cannot afford to hire ten associates or principals to source their deals for them. Here are some tips for developing a strong network with abundant proprietary deal flow.

1. **Scale your diligence capability:** Leverage outside industry experts, data engineers, and excited founders who amplify not just sourcing but also diligence.
2. **Leverage your extended network:** Build an extended team through advisors and venture partners.
3. **Turn portfolio founders into scouts:** Your most trusted referrals often come from the founders you've already backed. When you treat them like true partners—by being responsive, being helpful, and adding value—they become your strongest advocates and send the next generation of breakout talent your way before it hits the open market.

How to Make Everyone Feel Included in the Team

While building an early-stage fund, it's important to have an ecosystem of helpers. You can't do this alone, and in fact your portfolio companies will be worse off unless you make strategic friendships in the industry. There are ways to amplify your impact through building a feeling of camaraderie in your network.

Our Most Painful Lessons:
Hire the Hustlers

A well-incentivized rising star can do more for your fund than three full-time hires. We've seen junior team members move mountains when they're aligned with the mission and rewarded for the upside. Give them room to run—and the tools to win.

At Omni, we gave carried interest share to a lot of people. Not just full-time team members, but many people who materially helped us grow stronger: industry advisors, trusted referrers, even friends who opened doors in hard-to-crack geographies. If someone added value, they got a slice. Carry is bountiful in the early days, and the goodwill it buys lasts years. It turns loose networks into committed allies. And in a game where trust and reputation are everything, that kind of loyalty compounds faster than capital ever will.

Picking the Right LPs

Diversification of a fund's LP base is important. Relying solely on one geographical area can be a bad thing. Having too much ownership from a single strategic investor is also a bad thing.

Our Most Painful Lessons:
Diversify Your LP Base Early

We've seen friends struggle to raise Fund III or IV because all their LPs were from one region. Start diversifying early. The first LP in a new geography is the hardest to land, but once you do, they open doors you can't.

Here is a breakdown of the types of LPs and what they each care about and can offer you.

Institutional investors

These are the largest and most influential LPs, often with billions in assets:

- Pension funds—large pools of retirement savings looking for long-term returns (e.g., CalPERS, Ontario Teachers' Pension Plan)
- Endowments and foundations—nonprofits with capital to preserve and grow; known for early VC allocations (e.g., Yale, Stanford)
- Sovereign wealth funds—government-owned investment vehicles (e.g., Singapore's GIC, Abu Dhabi's Mubadala)
- Insurance companies—investing to meet long-term liability needs
- Funds of funds—LPs that pool capital from multiple investors and invest in a portfolio of VC funds

High-net-worth individuals and family offices

These LPs tend to be more flexible and risk tolerant than institutions:

- Angels or entrepreneurs—former founders or executives who invest in VC funds to get exposure to broader deal flow
- Family offices—wealth management firms dedicated and belonging to particular wealthy families. These can be traditional or entrepreneurial (like the Pritzkers or the Waltons, for instance).

Our Most Painful Lessons:
Fake Family Offices Are Everywhere

Here's a painful truth: 90 percent of the "family offices" that advertise themselves or attend family-office-branded events aren't real. We wasted too much time chasing ghosts. Real capital is quieter. It doesn't show up with glossy decks or generic intros. It comes through trusted networks. If someone's LinkedIn page touts "family office" in their title, take that as a red flag, not a credential.

Corporates (corporate LPs or CVCs)

These companies invest strategically to gain access to innovation. They may be motivated by financial returns, strategic insights, or the M&A pipeline. Examples include Google, Intel Capital, and Salesforce Ventures, which often invest both directly in start-ups and indirectly through VC funds.

Government-backed entities

These are public institutions that invest in VC to spur innovation or economic development.

- Development finance institutions (DFIs)—such as International Finance Corporation (IFC), the Business Development Bank of Canada (BDC), and the European Investment Fund
- Small business investment companies (SBICs) in the US—venture funds licensed by the Small Business Administration (SBA) to receive government leverage

University endowments

These are often pioneers in alternative investments like VC, but they are unlikely to invest in a Fund I unless they have an emerging manager program. Vanderbilt, Yale, Harvard, Stanford, and MIT are notable examples. Others include early backers of iconic funds like Sequoia or Benchmark.

Other VC funds (via funds of funds or re-ups)

Some VC firms act as LPs in other funds to get access to early-stage deals or diversify exposure. This is common in emerging GPs looking to create strategic network effects.

Our Most Painful Lessons:
Aligning with LP Interests

Some LPs, such as CVCs, aren't focused primarily on financial returns but still have their place in a healthy venture ecosystem. The key to alignment is figuring out what an LP is seeking and how you can help them get it.

Before pitching, listen to what they need and why.

Building a Strong Fundraising Network

A main priority of an emerging GP is to grow their network, particularly in the area of fundraising. Great GPs work to expand what I call their surface area of opportunity. The more engagements they have and people they help, the more opportunities they'll be able to generate for everyone's benefit. Once you help others, they feel indebted to you and are likely to do what they can to return the favor.

As a GP with a particularly international LP base, I'm a big believer in growing your network by creating champions in different regions. Champions are people who are excited about what you're doing and seem to have aligned interests. Having these advocates all over the world as extensions of your network will help you expand your web's reach. I painstakingly built my network of relationships not only through meetings with investment firms but also by hosting happy hours, breakfasts, and dinners in major cities all over the world. Some of my best connections were made while drinking and belting out songs at karaoke bars at two in the morning. Showing up in person whenever possible is still the fastest way to build trust, especially in new markets. Travel is expensive, exhausting, and wildly effective.

How can you start adding people to your extended network after arriving in a new city? The key is to start small and take a grassroots approach. See if you can find someone in the area you know, like an ex-colleague or someone you met at a conference, and let them know you're in town. Could they become one of your champions or introduce you to someone who could? Are they intrinsically motivated by an interest you could align with? Would they be a good partner for collaboration? Sit down with them over drinks, listen to them closely when they talk about their goals, and think about how you can help them.

Our Most Painful Lessons:
Small Regional LPs Are Gateways

When someone brings a truly unique network—especially in an under-tapped region—they're more than a check. They're a strategic partner. Prioritize those LPs at first. Their value isn't just capital—it's access.

While hustling to network and fundraise, you'll ideally end up in all sorts of meetings with different kinds of people. I've found that while most have good intentions, there are fraudsters in the world who are willing to waste your time. It's important to have boundaries around how you engage during first meetings. As you listen to the other person and figure out whether your interests align, try to do this quickly and keep the initial conversation minimal. After the meeting, keep that person updated on the status of your fund. Keeping yourself top of mind, even just through quarterly updates, keeps the door open for reciprocity in the future. You never know how some random person you met two or three years ago might be able to circle back and help you. Treat everyone as if they could become an LP at some point. Keep them in the loop and get them talking about you in their conversations with other people.

Having champions who believe in you mentioning your name to others over coffee or dinner is invaluable.

Our Most Painful Lessons:
Never Fully Count Anyone Out

Don't disqualify someone because you don't see how they can provide immediate value. LP momentum is a game of timing. Most want to see heat before they jump in. Until they unsubscribe from your updates, they're still in play. Keep showing up with progress, and when the timing's right, they'll remember you.

As you fundraise, it's important to have a well-defined exit strategy. As we covered while talking about DPI and IRR, VCs and start-ups have different options here. GPs should have an understanding of how their industry, location, fundraising stage, and legal obligations can affect the choices they negotiate with start-ups. If your industry is dominated by IPOs, for example, that's an important consideration. If IPOs aren't the norm, you'll have to come up with other ways to close your portfolio. Different industries have different exit potentials. Some might rely more heavily on M&As. Always do your homework and get clear on the options available to you.

The final point I want to share here is that fundraising takes time. VC funds are a blind pool, meaning your allocators have no idea which companies you'll be investing in. This requires huge amounts of trust on their part. The best way to deal with this is by focusing on these relationships themselves. Slow down and build personal rapport before diving into the nitty-gritty details of your fund. Talk about your fund in ways that differentiate you from others, and follow up by creating a series of touchpoints over an extended period of time. This gives prospective LPs time to build confidence in you and look over the materials you share. Each person will have a different level of risk tolerance, wanting to invest in your fund earlier, later, or not at all. Keep your head up by preparing yourself for the long game. The entire process of closing capital, for a typical emerging fund, can take anywhere from eighteen to twenty-four months on average. This is a marathon, not a sprint.

The venture market's maturation has not killed alpha; it has merely raised the bar for focus. As emerging technologies—AI, quantum, synthetic biology, space, climate—reshape trillion-dollar industries, the edge belongs to VCs who already live on that frontier. Identify your niche, master its nuance, and the next chapter is yours.

Can Generalists Be Niche?

GPs of emerging funds have unlimited options to choose from to differentiate themselves and grab the attention of specific types of founders. As this is the case, it's worth asking whether highly diversified VCs that write high volumes of small checks could be considered a niche of their own. Peter Walker, head of insights at Carta, recently defended the spray-and-pray approach of VCs that follow this strategy:

> Some LPs believe that the core value of the VC is to pick the right companies—so if your portfolio is 2–3× bigger than most seed-stage VCs, it signals you don't know what you're looking for. Of course a counter point may be that if you take more swings, each individual bet can be less consensus. Easier to back the wildest ideas if the fund has 50 portcos vs 25 . . . After you do a little math full of assumptions, the diversified portfolio beats the concentrated one for MOIC (Multiple on Invested Capital) at every point along the curve except for the super stellar funds. Said differently—if you are a top 5% fund, concentration wins. If you're not, diversification wins.[11]

---

[11] Peter Walker, "Should VCs Be Open to the Spray & Pray Approach?" LinkedIn, last updated July 2025,

In this book, I define niche as the three-pronged combination of mastery, focus, and network that correlates with alpha generation. If a fund is high in focus and network, those prongs can compensate for low mastery. Experience has taught me it is indeed possible for such funds to stand out. Hustle Fund, which invests in "hilariously early start-ups," writes over two hundred small checks to start-ups per year. While the firm isn't specifically technical, its team has managed to break through the noise with a loud and proud social media presence packed with educational content for founders. This illustrates the point that a strong platform is crucial for success in today's bloated VC market, and this is true whether a fund chooses to specialize or not.

The wisdom you gain as an LP will help you as you make the transition to fund manager if the idea of becoming a GP appeals to you. Prioritizing relationship building and becoming incredibly disciplined are the biggest steps you can take to get your emerging fund off the ground. As you do, you'll want to actively take steps to scale your business through an engaging platform. Next, we explore the many ways emerging GPs can work to establish their brand.

---

https://www.linkedin.com/posts/peterjameswalker_should-vcs-be-more-open-to-the-spray-pray-activity-7346213931729788928-WVv9/.

Chapter 9:
Executing on Your Niche Strategy

A Look Ahead (The TL;DR)

- Early-stage VCs can succeed in executing on their niche strategy by keeping up to date in their area of specialization, continuously fundraising, and leaning into the three prongs of the niche-VC trifecta.
- Start-up founders represent 90 percent of the value for investors. The best founders capitalize on their strengths and build well-rounded teams to compensate for their weaknesses.
- Founders lacking in relentless passion and fundamental business skills are a red flag and should be avoided.

When Berlin's Earlybird spun up its small $150 million Digital East Fund in 2014, few LPs cared about Central and Eastern Europe. That contrarian bet produced one of the biggest VC home runs of the decade: a 2,200× seed return on UiPath, helping the fund distribute $2.3 billion and vaulting it into the top tier of global performance tables. The lesson offered here is that geographic focus can work—if you really know the ground and arrive before capital crowds in.

Lessons Learned About Early-Stage Investing

Given that Omni Ventures is a niche early-stage fund, the main lessons I've learned over the years pertain to early-stage investing.

Keep Up to Date

Keep up to date with the latest information in your selected area of specialization. If needed, build your network to help you stay informed on the latest happenings.

Always Be ~~Fundraising~~ Networking

Always be fundraising, especially when you're not fundraising. A huge part of building a lasting firm with an enduring LP base that will grow with you is to build a diverse and scalable LP network. Just like founders should be meeting VCs early and before they are fundraising, GPs should be doing so continuously. For more detailed advice, check out my favorite book on this topic: *Fundraising* by Ryan Breslow.

Our Most Painful Lessons:
Let Others Tell Your Story

Fundraising is more powerful when someone else is telling your story. We learned this the hard way. Don't just pitch—rather, equip your early believers to hustle with you. Your first close LPs should be your best internal champions. Get them talking to their networks and yours. Their credibility is worth ten times more than your own voice.

The worst time to meet an LP is when you're already fundraising. The best time is when you don't need anything. Be curious. Ask questions. Listen. When there's no ask on the table, people are more honest—and those relationships tend to last longer.

Leverage the Customer-LP Flywheel

Some of our best deals came together when a strategic coinvestor was also a potential customer. That alignment can help start-ups land faster. Just make sure the customer LP isn't overbearing. You want a partner, not a parent.

Understand That Coleading Beats Over-Competing

At the early stage, coleading isn't a compromise—it's a strength. It gives the founder a more diversified cap table and lets you share the load. Save your elbows for later.

Lean Into Your Niche

As you execute on your niche strategy, keep in mind the three-pronged flywheel driving alpha in firms that specialize: mastery, focus, and network. Paying attention to each aspect of this niche-VC trifecta will enable you to spend your time and energy effectively as your fund grows.

A professional investor can expect to invest in twenty to one hundred start-ups every three years. To do so, you'll need to take the time to sincerely evaluate at least ten times that number, if not more, which amounts to at least two hundred companies. How can you get that number of deal flow as a beginner? The answer is to lean into your niche as much as possible. Build a brand around yourself and your firm. This involves figuring out ways to showcase yourself and scale your visibility as much as possible. As a GP, you're the face of your company and need to get people to perceive you as a thought leader. Any public appearances you can make in relation to your niche will be greatly beneficial to your brand. If your fund's focus is, say, underrepresented founders, you should be speaking up about that online and trying to get on stages to spread your message. If people don't seem to be listening, find out how to adjust your pitch so it's important and topical enough to grab their attention.

Connect with Your Audience

After you've honed your message for your niche, try to find people who are willing to generate momentum with you. They could be conference organizers, online groups, other emerging GPs, or LPs in the venture scene. Connect online and in person with diligence and consistency. Provide fresh, authentic content to your growing audience on a regular basis. Use YouTube, LinkedIn, X, TikTok, and any other social media outlet suitable for your brand.

Our Most Painful Lessons:
Brands Build Trust

A strong brand is more than visibility. It earns you trust before anyone meets you. It's why a founder replies to your cold email, why an LP takes the first meeting, and why people assume you belong in the room. In a world where everyone is selling access, brand is one of the most scalable ways to accelerate trust and signal value before you even speak. Spend time building this well.

The best niche VCs don't actually need to search for start-ups to fund. Deal flow comes inbound to them thanks to how well they're known in their specific field. Once you've established yourself as a thought leader in your field, both allocators and founders will view you as credible before you've even spoken in person. They'll come to you for advice rather than money, which is the ultimate way to build rapport with founders and ensure you have access to the hottest deals—a huge honor. The ultimate level of achievement is to create a platform that works, in effect, as a self-propagating deal-flow engine.

Back the Right People, Not Just the Right Ideas

Learning the traits, attitudes, and practices of a good founder is a huge part of managing a successful firm. If you manage a niche pre-seed fund like mine, the founders themselves represent 90 percent of the value for your investors. They are the single most important determinant of whether their company scales or fails.

**Our Most Painful Lessons:
Founders over Ideas**

The best founders make you believe. They can rally talent, inspire investors, and pivot when needed. At the early stage, the idea is secondary. You're investing in the spark, not the sketch.

Over the years, I've discovered the best founders know their weaknesses and build synergistic teams of people that can compensate for them. A big ego in want of an autocracy will sink a fledgling start-up in the fast-changing world of venture. You can't leverage your past experience for long in an industry in which the new frontiers of tech move at a rapid pace. The best founders know this and make an effort to always be learning. They're dynamic and able to embrace change. When they notice areas in which they fall short, they hire helpers to fill in those gaps for them, resulting in a highly skilled, well-rounded team.

I talked before about how exceptional GPs are driven by a sense of relentless passion for their field. The same is true for founders. These days, I've learned the best founders out there are self-taught. As GPs, we can teach founders and give them advice, but we can't make up for a lack of passion for their business. We can't equip them with the kind of energetic personality needed to fundraise effectively or handle business communications.

**Our Most Painful Lessons:
Balance Beats Brilliance**

The best founding teams aren't always the flashiest—they're the most complementary. One builds, one sells. Or one learns both. What matters most is self-awareness and chemistry.

Founders lacking in these basic qualities, who lack a cofounder to make up for these functional gaps, are a red flag and should be avoided. These aren't people who will be up until midnight updating their pitch deck before a meeting. They won't be jumping out of bed at six in the morning with a fresh idea they can't wait to put to use. That sense of relentlessness is far more important than whatever product they're looking to sell at their start-up. That's what it takes to win this game.

Chapter 10:
Building a Lasting Firm

A Look Ahead (The TL;DR)

- Emerging GPs must create scalable platforms for their brand if they're going to survive in today's overcrowded venture market.
- Specialization is a fantastic way to drive venture returns, but niche sectors inherently have capacity constraints. There's only so much capital that can be deployed before valuations inflate and returns erode.
- Building specialized infrastructures and enforcing internal constraints can help GPs avoid the common traps of style drift and overextension.

Investors who go on to start their own funds must create a scalable platform for their brand if they're going to survive in venture's overcrowded market. While building their team, they must balance the need for deal flow with their focus on their niche.

Don't Bloat

Specialization is a fantastic way to drive venture returns, but niche sectors inherently have capacity constraints. There's only so much capital that can be productively deployed before valuations inflate and returns erode. Smaller, specialized funds are naturally suited to grasp these opportunities because they can pursue concentrated, high-conviction bets without distorting the market. By contrast, mega-funds are structurally unable to target niche spaces. Deploying large sums would quickly oversaturate the ecosystem, driving up prices and compressing potential alpha.

Austin Venture's $1.5 billion misstep offers a cautionary tale of the threat of bloating a niche fund. In 2000, this respected firm raised a massive $1.5 billion fund—only the sixth VC fund ever to top $1 billion at the time. One of the firm's investors, Kevin Lalande, later revealed that this "bigger is better" move was a grave mistake. The gargantuan fund simply couldn't deploy capital as effectively. It had to spread bets wider and later, diluting the performance that Austin Ventures had achieved with smaller funds. Despite a strong brand and talent, the $1.5 billion fund failed to keep up its returns. Lalande noted that the venture industry repeatedly forgets this lesson: Raising a huge fund is often *counterproductive*. "Once you have a fund of sufficient size, if it's much bigger, it gets statistically more difficult to generate great returns," he explained.[12] In fact, only four of the thirty large funds over $400 million in one LP's study beat the stock market by a meaningful margin. This cautionary tale provides evidence that oversizing a fund can undermine performance, whereas staying right sized and focused keeps a venture firm hungry to excel.

How to Scale Without Bloating

Lux Capital grew from approximately $100 million in early-stage vehicles in the 2000s to greater than $1 billion in multistage flagships while staying laser focused on frontier science and deep tech. The firm accomplished this by executing on the following strategies:

- Carving subsector swim lanes (space, computation-driven biology, nuclear) for each partner
- Using its Lux R&D Fellows network (300+ PhDs) as an

---

[12] Sam Blum, "VC Funding in 2024: High-Profile Departures, Layoffs, and a Glut of Investors Struggling to Generate Returns," *Inc.com*, January 26, 2024, https://www.inc.com/sam-blum/venture-capital-funds-are-navigating-a-tough-start-to-2024.html.

evergreen proprietary-sourcing engine

- Maintaining ten to twelve core partners with the belief that platform talent scales diligence, not investment committee head count
- Capping its flagship fund at $1.15 billion and handling larger bites via coinvestors or SPVs to avoid style drift

Another success story is SOSV, currently valued at $1.5 billion AUM across accelerator-centric funds such as IndieBio, HAX, dlab, and Orbit Startups. These are the main aspects of their platform:

- Sector-specific accelerators create high-volume, proprietary deal flow in biotech, climate hardware, and Web3.
- Each accelerator partner signs less than 0.5 percent of all inbound applications—the edge is safeguarded by extreme filtering.
- Their follow-on pool (select fund) lets SOSV double down without bloating accelerator economics.
- Their internal Founder Hunger Score ties partner compensation to DPI, keeping incentives aligned.

Together, Lux and SOSV demonstrate that it's possible to scale venture platforms without sacrificing focus or discipline. By building specialized infrastructures—from Lux's deep-tech sourcing engine to SOSV's vertical accelerators—and enforcing internal constraints on fund size, team structure, and partner incentives, both firms have avoided the common traps of style drift and overextension. Their models prove that with clear swim lanes, smart platform design, and a commitment to alignment, scaling up doesn't have to mean bloating out.

Conclusion:
The Future of Venture

Throughout this book, we've explored the importance of each prong of the niche-VC flywheel associated with alpha: mastery, focus, and network. These factors coalesce to set small, early-stage VCs up for success in their areas of specialization. Mastery enables GPs to hone their funds' focus. Focus differentiates the brand and amplifies its signal, deepening the fund's network. Network expands the area where mastery can compound, connecting each spoke of the flywheel. At Omni Ventures, as GPs with years of experience in niche VC who managed to thrive throughout one of the industry's worst extinction events, we plan to continue going all in on specialization for the foreseeable future in our quest for alpha.

While the current wave of niche venture capital may be the cutting edge today, we must keep in mind that the frontier is always moving. What feels novel and visionary now will inevitably become the norm—or even obsolete—as markets, technologies, and founders evolve. Yet within this constant flux lies opportunity: The very transience of niche creates a window of alpha for those bold enough to seek it. Venture has never been static, and it is this restless dynamic that continues to reward those who are willing to adapt. The next frontier is already forming, and we must remain ready to meet it.

Glossary

Accelerator

Definition: A structured program that invests at a company's earliest stage in exchange for equity and provides mentorship, resources, and networking opportunities.

Why it matters: Early guidance and access to investor and customer networks can set the trajectory for a start-up's growth and significantly shorten the time it takes to reach product-market fit.

In this book: Accelerators are described as launchpads that, when well chosen, can compress years of progress into months for founders operating in a niche.

Afore Capital

Definition: A venture firm specializing in the pre-seed stage, focusing exclusively on being the first institutional investor in start-ups.

Why it matters: Specializing in an overlooked stage can create strong brand recognition among founders and improve access to the most promising opportunities before competition sets in.

In this book: Afore Capital is used as an example of how stage specialization can build defensible advantages and produce outsized results.

Alpha

Definition: The amount by which an investment outperforms its benchmark, such as a stock index or peer group.

Why it matters: Limited partners (LPs) invest in venture capital with the expectation of generating alpha, since it represents returns driven by unique insight and skill rather than general market movements.

In this book: Alpha is defined as the product of mastery, focus, and network, positioning it as something a disciplined general partner (GP) can intentionally create rather than leave to chance.

Anchor LP (Anchor Investor)

Definition: An anchor limited partner (LP) is a large, early investor whose commitment to a fund signals credibility and reduces perceived risk for other potential LPs.

Why it matters: Anchors often provide the momentum that accelerates a fundraise, opening the door for other investors to join more quickly.

In this book: Anchor LPs are treated as critical early wins that can change the psychology of a raise, creating urgency and social proof in the LP community.

AUM (Assets Under Management)

Definition: The total value of capital a fund manages on behalf of its investors, including both invested and uninvested amounts.

Why it matters: AUM directly influences management fees, which can in turn shape general partner (GP) incentives and the scale of the fund's operations.

In this book: The discussion warns against allowing AUM to grow beyond what the GP's niche can absorb, as doing so can dilute focus and returns.

Austin Ventures

Definition: A once-prominent venture capital firm that raised one of the earliest billion-dollar funds in 2000.

Why it matters: The size of a fund can limit its ability to deploy capital effectively in its chosen niche, which can lead to weaker returns.

In this book: Austin Ventures is presented as a cautionary tale about oversizing a niche fund and losing the edge that smaller scale provides.

Benchmark Capital

Definition: A venture firm known for its intentionally small fund sizes and concentrated, high-conviction early investments.

Why it matters: Maintaining discipline on fund size and focus can allow a firm to compete successfully against much larger rivals.

In this book: Benchmark Capital is used as a model of how staying small and committed to a thesis can generate outsized returns.

Beta

Definition: A measure of how much an investment's returns move in line with the overall market.

Why it matters: High beta means that returns are being driven by macroeconomic conditions rather than the unique skill or insight of the general partner (GP).

In this book: Beta is used as a counterpoint to alpha and as a warning sign that a strategy may be drifting into market-tracking behavior.

Beta-Like Returns

Definition: Outcomes that closely match general market performance, with little to no outperformance.

Why it matters: These returns are common in later-stage mega-funds, where upside potential is capped but downside is also limited.

In this book: Beta-like returns are presented as evidence that some funds have shifted from hunting for alpha to managing large pools of capital for predictability.

Burn Rate

Definition: The amount of cash a start-up spends each month.

Why it matters: Burn rate determines runway, which is the amount of time a company has before it must raise more capital or become profitable.

In this book: Burn rate is discussed as a signal of founder discipline and operational control, especially in capital-constrained markets.

CAC (Customer Acquisition Cost)

Definition: The average cost of acquiring a paying customer, factoring in all marketing and sales expenses.

Why it matters: A sustainable business model depends on a CAC that is low enough relative to the revenue each customer generates.

In this book: CAC is always considered alongside lifetime value (LTV) and payback periods to ensure the economics are viable.

Cap Table

Definition: A document that shows the ownership breakdown of a company, listing all shareholders and their respective stakes.

Why it matters: Clean and balanced cap tables make future fundraising rounds easier and more attractive to new investors.

In this book: Cap tables are emphasized as part of maintaining ownership discipline and protecting the fund's ability to realize strong returns.

Capital Call

Definition: A request by a general partner (GP) for limited partners (LPs) to transfer a portion of their committed capital into the fund for investment or operational expenses.

Why it matters: The timing and frequency of capital calls can affect both the GP's ability to act quickly and the LP's liquidity management.

In this book: Capital calls are described as an operational tool that reflects the GP's professionalism and ability to manage fund cash flow.

Carried Interest (Carry)

Definition: The general partner's (GP's) share of a fund's profits, typically around 20 percent, earned after returning the limited partners' (LPs') original capital.

Why it matters: Carry aligns GP incentives with LP outcomes, motivating managers to maximize net returns.

In this book: Carry is presented as meaningful only when it translates into realized distributed to paid-in capital (DPI), rather than just paper gains.

Check Size

Definition: The amount of capital a fund invests in a single deal.

Why it matters: It directly influences ownership levels, portfolio construction, and potential fund-level returns.

In this book: Check size is sized deliberately to align with the general partner's (GP's) niche strategy and maintain the math needed to achieve target multiples.

Coinvestment Rights

Definition: Giving limited partners (LPs) or general partners (GPs) the option to invest additional capital directly in a portfolio company alongside the main fund.

Why it matters: This allows LPs to increase their exposure to top-performing companies without paying additional management fees or carry.

In this book: Coinvestment rights are framed as a strategic tool for strengthening LP relationships and securing alignment.

CVC (Corporate Venture Capital)

Definition: Venture investing conducted by a corporation, usually from its balance sheet.

Why it matters: CVCs can offer strategic advantages such as distribution, partnerships, and market validation, though their motivations may differ from purely financial investors.

In this book: CVCs are portrayed as both valuable partners and potential competitors, depending on strategic fit with the general partner's (GP's) portfolio.

Decacorn

Definition: A private company valued at $10 billion or more.

Why it matters: These valuations often occur at later stages when upside potential is reduced, making them less accessible to early-stage general partners (GPs).

In this book: Decacorns are presented as by-products of compounding niche advantages, rather than primary investment targets.

DFI (Development Finance Institution)

Definition: A publicly backed investment organization that funds private-sector projects to promote economic growth, often in developing markets.

Why it matters: DFIs can provide significant limited partner (LP) capital and open doors to international opportunities, particularly for funds with impact or geographic theses.

In this book: DFIs are discussed as valuable but specialized LP partners, especially for funds aligned with developmental or emerging-market goals.

Dilution

Definition: Occurs when a company issues new shares, reducing the ownership percentage of existing shareholders.

Why it matters: In venture capital, excessive dilution can erode a fund's ability to achieve strong returns, even if the company succeeds.

In this book: Ownership discipline is emphasized as a critical safeguard against value loss from dilution.

DPI (Distributed to Paid-In Capital)

Definition: Measures the amount of cash or stock returned to limited partners (LPs) relative to the total amount of capital they contributed.

Why it matters: It reflects realized returns, which are the ultimate measure of a fund's success.

In this book: DPI is treated as the primary scoreboard, valued more highly than unrealized or paper gains.

Early DPI Strategy

Definition: A deliberate plan to return capital to limited partners (LPs) sooner than the typical venture timeline, often via early M&A exits or secondary sales.

Why it matters: Early distributions improve internal rate of return (IRR) and build LP confidence in the general partner's (GP's) execution.

In this book: Early DPI strategies are presented as a hallmark of professional, LP-aligned fund management.

Emerging Manager

Definition: A newer or smaller general partner (GP) who is often running their first, second, or third fund.

Why it matters: Many of the best-performing funds historically have come from emerging managers who can be nimble and differentiated.

In this book: Emerging managers are positioned as fertile ground for limited partners (LPs) seeking uncorrelated alpha in a crowded market.

Endowment

Definition: A long-term investment pool, often associated with universities or nonprofits, that seeks to preserve and grow capital while funding institutional goals.

Why it matters: Endowments were early adopters of venture capital and remain influential limited partners (LPs).

In this book: Endowments are portrayed as ideal partners for specialist managers due to their patience and return expectations.

ERISA

Definition: The Employee Retirement Income Security Act of 1974, a US law that allows pension funds to invest in higher-risk assets like venture capital.

Why it matters: ERISA opened up a massive pool of institutional capital to the venture industry, accelerating its growth.

In this book: ERISA's impact is used to illustrate how regulatory changes can transform limited partner (LP) behavior and fund formation.

Exit

Definition: A liquidity event where investors convert their ownership in a company into cash or tradable stock, typically through an initial public offering (IPO), merger, acquisition, or secondary sale.

Why it matters: Exits are the moment when paper gains become realized returns, directly affecting distributed to paid-in capital (DPI) and a fund's performance record.

In this book: Exits are framed as outcomes that should be deliberately designed for in a fund's strategy, rather than left to chance.

Extinction Event

Definition: A severe market downturn that eliminates a large portion of active venture funds, especially those without a differentiated strategy.

Why it matters: These events reset valuations, purge weaker players, and often create opportunities for well-positioned funds to outperform in the recovery.

In this book: Surviving an extinction event is portrayed as validation of a general partner's (GP's) discipline, resilience, and niche advantage.

Family Office

Definition: A private investment firm that manages the wealth of a single family, often across multiple asset classes.

Why it matters: Family offices tend to have flexible investment mandates and can make quicker decisions than large institutions.

In this book: They are depicted as high-signal early limited partners (LPs) whose belief can help catalyze other investors to join a fund.

Fee Load

Definition: The total amount of management fees collected over the life of a fund and their effect on net returns to limited partners (LPs).

Why it matters: High fee loads can significantly erode LP returns, especially if a fund's assets under management (AUM) grow beyond what its strategy can effectively deploy.

In this book: Fee discipline is emphasized as part of staying focused on craft rather than building an "AUM empire."

Follow-On Capital (Select Fund)

Definition: Money reserved or raised to reinvest in the fund's strongest portfolio companies in later rounds.

Why it matters: Maintaining ownership in winners can dramatically increase a fund's overall returns.

In this book: Follow-on strategies are presented as a critical component of portfolio design and ownership discipline.

Founder Hunger Score

Definition: A metric used by SOSV that ties partner compensation to realized returns (distributed to paid-in capital, or DPI) rather than just paper gains.

Why it matters: Linking pay to realized outcomes aligns the general partner (GP's) incentives more closely with limited partner (LP) interests.

In this book: It's cited as an example of how incentives can be structured to keep focus on tangible performance.

Frontier Tech

Definition: Technologies at the cutting edge of science and engineering, such as AI, climate technology, space systems, and biotech.

Why it matters: These areas often carry high technical risk but also the potential for transformational, category-defining companies.

In this book: Frontier tech is positioned as fertile ground for niche VCs who possess true technical mastery.

Fund of Funds (FOF)

Definition: An investment vehicle that allocates capital across multiple venture capital funds rather than directly into start-ups.

Why it matters: FOFs provide diversification for their investors and can serve as important capital sources for emerging managers.

In this book: They are described as valuable early limited partners (LPs) who can also act as references and connectors within the LP ecosystem.

Generalist VC

Definition: A venture fund that invests across a broad range of sectors, stages, and geographies.

Why it matters: Generalist strategies can lack focus, making it harder to stand out in a crowded market.

In this book: A generalist fund can still be considered niche if it remains small, disciplined, and tightly aligned with a distinctive brand.

GP (General Partner)

Definition: The individual or group responsible for managing a venture capital fund, making investment decisions, and supporting portfolio companies.

Why it matters: The GP's judgment, network, and ability to execute are the primary drivers of a fund's performance.

In this book: GPs are framed as craftspeople who must resist the temptation to become asset gatherers.

GP Commit (Skin in the Game)

Definition: The amount of personal capital a general partner invests in their own fund.

Why it matters: A meaningful GP commit signals alignment with limited partners (LPs), as the GP has their own money at risk.

In this book: GP commit is presented as a tell for integrity, conviction, and belief in the strategy.

Gross vs. Net Returns

Definition: Gross returns are calculated before fees and carried interest, while net returns are what limited partners (LPs) actually receive after all costs.

Why it matters: Net returns are the true measure of what matters to LPs; gross numbers can be misleading.

In this book: All performance discussions are based on net returns to LPs to reflect real-world results.

Hilltop Advantage

Definition: The unique perspective gained from deep domain mastery, allowing an investor to see opportunities others miss.

Why it matters: This vantage point can reveal nonobvious patterns and connections that lead to better investments.

In this book: Hilltop advantage is presented as a central pillar of the mastery component of the niche flywheel.

Hustle Fund

Definition: A high-volume pre-seed venture firm known for its strong public presence and educational content for founders.

Why it matters: A large platform and high visibility can help a fund stand out even without a narrow sector focus.

In this book: Hustle Fund is used as an example of how brand and platform can themselves serve as a form of niche positioning.

IPO (Initial Public Offering)

Definition: The process of a private company listing its shares on a public stock exchange, making them available for public trading.

Why it matters: IPOs can provide significant liquidity and establish a market valuation for the company.

In this book: IPOs are acknowledged as one path to exit, though M&A are often more common for early-stage investments.

IRR (Internal Rate of Return)

Definition: The annualized rate of return for an investment, taking into account the timing of cash flows.

Why it matters: Early distributions can significantly boost IRR, making it a key performance metric for limited partners (LPs).

In this book: IRR is managed intentionally through strategies like early distributed to paid-in capital (DPI) and disciplined reserve allocation.

LTV (Lifetime Value)

Definition: The total net revenue a company expects to earn from a customer over the entire relationship.

Why it matters: Comparing LTV to customer acquisition cost (CAC) helps assess a business's profitability and scalability.

In this book: LTV is evaluated alongside retention rates and payback periods to ensure long-term economic health.

Lux Capital

Definition: A venture firm specializing in deep-tech and frontier science companies.

Why it matters: Lux demonstrates that it is possible to scale a platform while maintaining a sharp focus on a specific niche.

In this book: Lux Capital is used as an example of disciplined growth without style drift.

Management Fee

Definition: The annual percentage of committed capital paid to a general partner (GP) to cover the costs of running a fund.

Why it matters: While necessary to operate the fund, excessive fees can distort incentives toward raising larger funds.

In this book: Management fees are discussed in the context of right-sizing to support the craft rather than fund empire-building.

Mastery × Focus × Network

Definition: The flywheel framework presented in this book as the core engine of niche VC success.

Why it matters: When combined, these elements create compounding advantages that improve deal flow, founder access, and portfolio performance.

In this book: This formula is the central thesis, illustrating how disciplined specialization drives alpha.

MOIC (Multiple on Invested Capital)

Definition: The ratio of total value (realized and unrealized) to the amount of capital invested.

Why it matters: MOIC offers a simple snapshot of investment performance, though it must be viewed alongside distributed to paid-in capital (DPI) and total value to paid-in (TVPI).

In this book: MOIC is interpreted in context with other metrics to provide a complete picture of fund performance.

M&A (Mergers and Acquisitions)

Definition: The consolidation of companies through various types of financial transactions, including mergers and acquisitions.

Why it matters: M&A is a common and often faster route to liquidity than an initial public offering (IPO), especially for early-stage companies.

In this book: M&A is discussed as an intentional part of many exit strategies, particularly in specialized markets.

Niche VC (Specialist VC)

Definition: A venture fund with a clearly defined focus area in which it can build a repeatable edge.

Why it matters: Specialization can improve access to deals, pricing, and the ability to add post-investment value.

In this book: Niche is defined as much by approach and execution as by sector or stage focus.

Notation Capital

Definition: A New York–based pre-seed venture firm that focuses on writing the very first institutional checks to start-ups.

Why it matters: Clear and consistent positioning in a specific stage can attract the right founders and deals.

In this book: Notation Capital is used as a case study of specialization as a form of branding and competitive advantage.

Ownership Drift

Definition: Occurs when a fund falls below its target ownership in a company, often due to overpaying or dilution.

Why it matters: This can undermine a fund's ability to achieve strong returns, even if the company performs well.

In this book: Ownership discipline is emphasized as a nonnegotiable part of a fund's long-term success.

Paper Markup

Definition: An increase in the valuation of a portfolio company based on a new funding round, without any realized liquidity.

Why it matters: While paper markups can signal progress, they do not translate to actual returns until realized.

In this book: Paper markups are seen as secondary to realized distributed to paid-in capital (DPI) and are not treated as indicators of true success.

Pension Fund

Definition: A large pool of retirement savings managed to provide income for retirees.

Why it matters: Pension funds are major institutional limited partners (LPs), though they are often conservative in backing emerging managers.

In this book: Pension funds are described as late-cycle adopters who can become important partners once a GP has established a track record.

Platform (at a VC)

Definition: The set of resources a fund provides to portfolio companies beyond capital, such as recruiting, go-to-market help, or community building.

Why it matters: A strong platform can differentiate a fund in competitive deal processes and improve portfolio outcomes.

In this book: Platforms are evaluated on how well they align with and reinforce the fund's niche.

PMF (Product-Market Fit)

Definition: Occurs when a product satisfies strong market demand, usually proven by retention and growth metrics.

Why it matters: Achieving PMF is often a key inflection point for scaling investment.

In this book: PMF is discussed in terms of evidence-based validation rather than founder anecdotes.

Pre-Seed

Definition: The earliest institutional funding stage, typically before a product has launched.

Why it matters: This stage often offers clean ownership and the opportunity to shape a company's direction from inception.

In this book: Pre-seed is highlighted as an ideal point for niche funds to secure early, high-quality positions.

Pro Rata Rights

Definition: Allow an investor to maintain their ownership percentage in a company by participating in future funding rounds.

Why it matters: They are essential for compounding returns in the fund's top-performing companies.

In this book: Pro rata rights are supported by disciplined reserves planning to ensure they can be exercised when needed.

Proprietary Deal Flow

Definition: High-quality investment opportunities sourced through a fund's network rather than public channels or auctions.

Why it matters: Proprietary sourcing often leads to better pricing and stronger founder relationships.

In this book: It is portrayed as the natural output of mastery and network working together in the niche flywheel.

Reserves

Definition: Capital set aside for follow-on investments in existing portfolio companies.

Why it matters: Proper reserves management ensures the fund can defend its ownership and support its best performers.

In this book: Reserves are explicitly modeled in portfolio construction to align with the fund's thesis and stage focus.

Runway

Definition: The amount of time a company can operate before it runs out of cash, based on its current burn rate.

Why it matters: Understanding runway is critical for anticipating funding needs and growth pacing.

In this book: Runway management is viewed as a sign of founder discipline and strategic planning.

SaaS (Software as a Service)

Definition: A software delivery model in which applications are hosted online and accessed by customers via subscription.

Why it matters: SaaS businesses are valued for their predictable recurring revenue and often high margins.

In this book: SaaS is favored when its unit economics and retention rates are strong enough to support long-term profitability.

Scout

Definition: An operator, founder, or trusted contact who sources deals for a venture firm, sometimes with authority to invest small amounts or receive a share of carry.

Why it matters: Scouts can expand a fund's reach into founder networks and surface high-quality opportunities before they become competitive.

In this book: Scouts are used effectively when their incentives are aligned and their judgment complements the general partner's (GP's) niche.

Secondary Transaction

Definition: The sale of private company shares before a full liquidity event like an IPO or acquisition.

Why it matters: Secondaries can provide partial liquidity to early investors and founders, as well as early distributed to paid-in capital (DPI) for funds.

In this book: Secondary sales are framed as a tool for generating returns without abandoning the long-term potential of a portfolio company.

Sequoia Capital

Definition: One of the most prominent venture capital firms in the world, known for early investments in companies such as Apple, Google, and Airbnb.

Why it matters: Sequoia sets industry standards for brand building, founder support, and strategic scaling.

In this book: Sequoia serves as a benchmark for balancing scale with focus, a challenge many funds face as they grow.

Side Letter

Definition: An agreement that grants a limited partner (LP) special rights or terms outside the main limited partnership agreement, such as reduced fees or coinvestment access.

Why it matters: Side letters can attract key LPs but may create precedent for other investors.

In this book: They are used sparingly to maintain fairness across the LP base while meeting strategic needs.

Sovereign Wealth Fund

Definition: A state-owned investment pool that manages national savings for long-term objectives.

Why it matters: These funds can provide large, patient capital but often have strategic or policy-driven considerations.

In this book: They are considered valuable limited partners (LPs) when their mandates align with the fund's niche and time horizon.

Spray and Pray

Definition: A high-volume investing strategy in which a fund makes many small investments with the hope that a few will deliver outsized returns.

Why it matters: This approach relies heavily on the power law but often sacrifices ownership and engagement.

In this book: Spray and pray is rejected unless paired with a platform that can deliver real value to a large portfolio.

SPV (Special Purpose Vehicle)

Definition: A one-off legal entity created to invest in a single company or specific deal outside the main fund.

Why it matters: SPVs allow funds to participate in off-thesis or oversized opportunities without distorting their main fund strategy.

In this book: SPVs are used to protect focus while capturing upside in exceptional situations.

Support Drift

Definition: Occurs when a fund invests in companies it cannot materially help after the check is written.

Why it matters: This weakens the fund's brand and can reduce founder trust and portfolio success rates.

In this book: Support drift is avoided by ensuring each investment aligns with the fund's post-investment value-add capabilities.

TAM (Total Addressable Market)

Definition: The total revenue opportunity available if a product or service achieved 100 percent market share in its target market.

Why it matters: TAM helps investors gauge the potential scale and upside of a business.

In this book: TAM is validated using bottom-up analysis, not just top-down estimates, to ensure realistic growth assumptions.

Term Sheet

Definition: A nonbinding agreement outlining the key terms and conditions of a proposed investment.

Why it matters: It frames valuation, governance rights, and economic terms for final legal agreements.

In this book: Term sheets are seen as an expression of partnership intent, not just a transaction document.

Thesis Drift

Definition: Occurs when a fund expands beyond its stated investment focus, often chasing trends or reacting to market pressures.

Why it matters: Drift can erode a fund's edge, confuse its brand, and weaken performance.

In this book: Thesis drift is viewed as a slippery slope toward beta-like returns and loss of differentiation.

TVPI (Total Value to Paid-In)

Definition: The ratio of a fund's total value (distributed to paid-in capital, or DPI, plus remaining unrealized value) to the capital paid in by limited partners (LPs).

Why it matters: It shows the combined realized and unrealized performance of a fund.

In this book: TVPI is always interpreted alongside DPI and vintage context to give a full performance picture.

Unicorn

Definition: A privately held start-up valued at $1 billion or more.

Why it matters: Unicorns often symbolize outlier success in venture portfolios, though they may not always deliver proportional returns.

In this book: Unicorn status is treated as the by-product of a winning strategy, not the goal in itself.

Unit Economics

Definition: The profitability of serving a single customer or producing one unit of product.

Why it matters: Healthy unit economics are critical for sustainable growth and scalability.

In this book: Unit economics are stress-tested with retention data and payback periods, not just lifetime value (LTV) / customer acquisition cost (CAC) ratios.

VAS (Venture Arrogance Score)

Definition: A metric created by Josh Kopelman to test whether a fund's strategy requires capturing an unrealistic share of all market exits.

Why it matters: A high VAS indicates implausible expectations, serving as a red flag for limited partners (LPs).

In this book: VAS is presented as a simple sanity check that should be part of every fund's design process.

VC (Venture Capital)

Definition: A form of private equity investing that provides funding to high-growth start-ups in exchange for equity.

Why it matters: VC aims to capture outlier returns by backing companies that have the potential for exponential growth.

In this book: Venture capital is treated as a craft that works best when applied with niche specialization.

Vintage (Fund Vintage Year)

Definition: The year a fund begins making investments.

Why it matters: Vintage is used to compare funds of similar age and economic conditions, enabling fairer performance benchmarks.

In this book: Vintage context is always considered when interpreting distributed to paid-in capital (DPI), total value to paid-in (TVPI), and internal rate of return (IRR).

ZIRP (Zero-Interest-Rate Policy)

Definition: A monetary policy in which central banks keep interest rates near zero to stimulate borrowing and investment.

Why it matters: In the US, ZIRP periods from 2008 to 2015 and 2020 to 2022 flooded markets with cheap capital, inflating venture valuations and deal volume.

In this book: The end of ZIRP in 2022, marked by rapid rate hikes, is described as a turning point that triggered funding contractions and rewarded disciplined, niche-focused managers.

Printed in Great Britain
by Amazon

a0df5706-9492-479c-8334-8372af200886R01